Number Four: The Montague History of Oil Series

Early California Oil

A mirrow view of the Lakeview No. 1 at Midway in the huge sump built to contain the crude. *Pacific Oil World*

Early California Oil

A Photographic History, 1865–1940

By KENNY A. FRANKS

and PAUL F. LAMBERT

TEXAS A&M UNIVERSITY PRESS College Station

Library of Congress Cataloging in Publication Data

Franks, Kenny Arthur, 1945–
Early California oil.

(Montague history of oil series ; no. 4)
Bibliography: p.
Includes index.
1. Petroleum industry and trade—California—History.
I. Lambert, Paul F. II. Title. III. Series.
HD9567.C2F73 1985 338.2′7282′09794 84-40566
ISBN 0-89096-206-5

Manufactured in the United States of America
FIRST EDITION

For Bryan James Franks

Contents

Preface

CALIFORNIA is one of the most intriguing states in the Union. Beautiful beaches; spectacular snow-capped peaks; fertile, productive river valleys; massive verdant forests; expansive deserts; and rugged, sparse hills—all are found within the borders of the Golden State. The state's geography creates tremendous ranges of climate, with marked variations in rainfall and temperature. Californians brave the constant possibility of earthquakes, Pacific storms, mudslides, and forest fires to enjoy the physical glories and economic opportunities that their state has offered throughout its history.

California is as varied and colorful in its history as it is diverse and beautiful in its geography. Native Americans enjoyed the natural bounty of what would become California for centuries before the arrival of Western European and Russian explorers. The desirability of colonizing this region was quickly apparent. The Spanish conquistadores and padres were successful in this effort, and their influence on the state's architecture, place names, and culture remains evident to the present. The rise of the Bear Flag Republic, the discovery of gold at Sutter's Mill and the resultant rush to the region, the development of the new state's agriculture, the growth of large, sophisticated cities, the tides of immigration, and the evolution of the state's contemporary cosmopolitan society each has been the subject of numerous books and articles. Historians and novelists alike have written extensively about the Spanish era, the achievement of independence from Mexico, and the gold and silver mining booms of the nineteenth century. Each of these dramatic, romantic epochs certainly deserves the attention it has received, but another important and fascinating aspect of California's experience—a subject equally deserving of scholarly and public attention—has been largely ignored by state historians.

The history of the oil and gas industry in California generally receives scant attention in state history texts, yet the experiences of gambling wildcatters, enterprising entrepreneurs, and hard-working oil-field laborers are as exciting and kaleidoscopic as the history of any other part of the state's history. California oil and gas gave birth to bustling boom towns, tremendously stimulated the state's economy, led to the founding of corporations with worldwide significance, and played a major role in the nation's economic and historical development. Moreover, California oil men, responding to novel geological problems encountered in the state, developed techniques and technologies that have had international ramifications.

This work is a visual exploration of the growth and development of California's rich and colorful petroleum legacy from its beginning through the boom years of the first four decades of the twentieth century. California's petroleum history during this era can be divided into three periods. At first the region's numerous natural oil seeps were utilized by Native Americans and early Spanish settlers as pitch for canoes and ships, illumination, and medicine. Later, with the arrival of the Americans, the state's rich oil reserves took on a new meaning as continual new discoveries of crude took place. The birth of California's petroleum industry proper may be dated to 1865, with the first commercial sale of oil refined in the state (by the Stanford brothers) from a well in the state (on the Mattole River, in Humboldt County). At first, though, there was no great increase in markets, and despite the use of the oil as surfacing for roads, fuel for ships and railroads, and natural gas for lighting, the California petroleum industry languished. However, with the beginning of the twentieth century, the third period, the actual boom, was on. Contributing more than anything else to the beginning of the boom were the development of ocean-going tankers that opened vast new markets and the growing demand for gasoline created by the "Age of the Automobile."

Between 1887 and 1949, twenty-one of America's highest-producing twentieth-century oil fields were brought in in California. The fields, dates of discovery, and national rankings based on cumulative production as of 1949 were: Midway-Sunset (1901), 2nd; Long Beach (1921), 3rd; Santa Fe Springs (1919), 5th; Wilmington (1935), 7th; Huntington Beach (1920), 8th; Buena Vista (1909), 11th; Ventura Avenue (1916), 12th; Kettleman Hills fields (1928), 13th; Kern River (1899), 16th; Coalinga East (1901), 18th; Brea-Olinda (1889), 26th; Elk Hills (1919), 27th; Dominguez (1923), 34th; Inglewood (1924), 36th; Coalinga West (1901), 38th; West Coyote (1909), 39th; Montebello (1917), 45th; Coalinga Nose, or Coalinga East Extension (1938), 46th; Torrance (1922), 57th; McKittrick-Cymric (1887), 65th; and Orcutt, or Santa Maria (1902), 72nd. Total cumulative production from these fields between 1887 and the end of 1949 was an astounding 6,360,722,000 barrels of crude.

California ranked first among the states in oil production in 1903–1905, 1909–14, 1919, and 1923–26. The state was second in production in 1906–1908, 1915–18, 1920–23, 1927, 1929–32, and 1935–40 and third overall in 1928 and 1933–34. California was the only state in the union that was ranked by itself as an oil-producing region. Indeed, it was first among the oil-producing regions in 1906 and 1909–14.

We have drawn upon photographs to tell the colorful history and heritage of the California petroleum industry. These illustrations, combined with contemporary descriptions by many of those who participated in the California oil boom, present a graphic portrayal of the scenes and characters involved. Together they offer a unique view of California's second great mineral rush.

Acknowledgments

THE cooperation of many individuals, corporations, and institutions made the publication of *Early California Oil* possible. These people and their institutional or corporate affiliations include: Leslie Brooks, Leslie Brooks and Associates, Tulsa; Roxanne Burg, Atlantic Richfield Company, Los Angeles; Bonnie Chaikind, Getty Oil Company, Los Angeles; Andrea Clark, Oklahoma Historical Society, Oklahoma City; Peggy Coe, Oklahoma Department of Libraries, Oklahoma City; Richard Drew, American Petroleum Institute, Washington, D.C.; Bettye Ellison, California Room, Los Angeles Public Library; Karen Fite, Oklahoma Department of Libraries, Oklahoma City; Carol Guilliams, Oklahoma Department of Libraries, Oklahoma City; Mary Hardin, Oklahoma Department of Libraries, Oklahoma City; Manny Jimenez, Atlantic Richfield Company, Los Angeles; James O. Kemm, Oklahoma-Kansas Oil and Gas Association, Tulsa; Claren Kidd, Geology Library, University of Oklahoma, Norman; Kimberly Kirkpatrick, Oklahoma Historical Society, Oklahoma City; Julia Koehler, Oklahoma Department of Libraries, Oklahoma City; P. C. Lauinger, PennWell Publishing Company, Tulsa.

Also, Brita Mack, Huntington Library, San Marino; Frank Parisi, Getty Oil Company, Los Angeles; Betty Quirk, Standard Oil Company of California, San Francisco; Jack M. Rider, *Pacific Oil World*, Brea; Karen Sikkema, Union Oil Company of California, Los Angeles; Arthur O. Spaulding, Western Oil and Gas Association, Los Angeles; Fred Stanley, Oklahoma Historical Library, Oklahoma City; John Steiger, Cities Service Oil Company, Tulsa; Mary Sweeney, Oklahoma Department of Libraries, Oklahoma City; Bob Thomas, Standard Oil of California, San Francisco; Marie Tilson, Standard Oil of California, San Francisco; Charlesa Timmons, Oklahoma Historical Society, Oklahoma City; Otis Tobey, Union Oil Company of California, Los Angeles; and Steve Beleu, Oklahoma Department of Libraries, Oklahoma City.

Without the aid of these individuals—and their corporations and organizations—who gave readily of their knowledge of the California petroleum industry, this book would never have been completed. Finally, the research assistance, proofreading, encouragement, and understanding provided by our wives, Jayne White Franks and Judy K. Lambert, were vital to this effort.

Early California Oil

Introduction

PETROLEUM played an important role in the history of California long before the area became part of the United States. In 1543 Juan Cabrillo, a Portuguese sailing under the Spanish flag, recorded that the Indians along the Santa Barbara Channel used asphaltum to caulk their canoes. Cabrillo followed the Indians' example, using the substance to waterproof two ships. Another Spanish explorer reported in 1775 that within two leagues of the mission at San Luis Obispo there were "as many as eight springs of a bitumen" that the natives called *chapapote*, and which they used primarily for "caulking their small water craft, and to pitch the vases and pitchers which the women make for holding water." The Chumash Indians perfected a wooden canoe constructed of planks bound together by small ropes. They caulked the cracks and the binding holes with asphaltum. They also used the substance as an adhesive for affixing arrowheads to the shafts and for attaching bone mouthpieces to pipes.

Much of the oil used by the Indians came from the La Brea tar pits, located near Los Angeles. The heavy oil oozing from the ground also attracted many early oil men. In 1888, Lyman Stewart and Dan McFarland drilled a wildcat nearby, but it proved to be a duster. When W. W. Orcutt, the original organizer of the geological department of Union Oil of California, reexamined the area in 1901, he discovered "a vast mosaic of white bones" on the surface of a pool of asphalt—the skeleton of a giant ground sloth, a huge armored animal that had been extinct for millions of years. As paleontologists subsequently probed the La Brea tar pits, it became obvious that the heavy asphalt had trapped numerous prehistoric animals and, more important, had then perfectly preserved their skeletons. It was perhaps the richest paleontological find ever made.

In addition to the tar pits, the Indians also gathered lumps of bitumen from the beaches, where it accumulated naturally from the many submarine seeps off California's coast. In 1792 Captain George Vancouver of the British Navy recorded observing one of these huge slicks off the coast of Santa Barbara. According to Vancouver, the oil was so thick that the entire sea took on an iridescent hue.

The Spaniards who colonized California were well aware of the existence of oil seeps. However, they made little use of the oil, other than for caulking their ships. Motivated by zeal for converting the Indians to Catholicism, prevention of Russian colonization, and the acquisition of riches, the Spaniards ironically did not know that this land possessed vast stores of both yellow and black gold.

After the Mexican revolt against Spain—when California became a part of Mexico—traditional uses of bitumen continued, although some of the oil from seepages in Ventura County reportedly was processed for use in lamps in missions. But California's history was drastically altered with James W. Marshall's discovery of gold on January 24, 1848, at Sutter's Mill. Within three years, the annual value of California's gold production reached fifty-five million dollars. Thousands of Americans and other gold seekers poured to the West Coast in search of riches, rapidly changing California from a rural agricultural region to a bustling trading area. Inevitably, the Americans began to clamor for admission to the Union. In 1848, by the terms of the Treaty of Guadalupe Hidalgo, California was ceded to the United States. Nine years before Colonel Edwin L. Drake drilled America's first oil well in Pennsylvania, California was granted statehood as part of the Compromise of 1850.

Many of the later Californians had long been aware of the potential petroleum reserves of the region. Cris Danielson, an early-day pioneer, recalled that after he established his home near Coalinga, his family used to pack a picnic lunch and drive the family wagon to one of the nearby oil seeps. There they would spend the day collecting buckets of oil that they used as axle grease. In 1852 Colonel James Williamson reported a rich vein of bitumen on the Rancho Ojai, a few miles northeast of Ventura. The next year, 1853, William P. Blake, a geologist, reported "bituminous effusions" about three hundred miles south of San Francisco. The same deposit was recorded by Thomas Antisell in 1855.

As early as the 1850s, the Americans began to tap this plentiful natural resource. In 1851, George Dietz and Company was operating a camphene still in San Francisco. By 1855 several San Francisco streets were paved with asphalt. In 1856, still three years before Drake's well was drilled, Andreas Pico was distilling oil taken from a natural seep in northern Los Angeles County for use in lighting the local San Fernando Mission. In 1857 Charles Morrell, a San Francisco druggist, built a distilling plant near the maltha seepage at Carpinteria in Santa Barbara County. These efforts, though, resulted in only limited commercial success.

By the mid-1850s, however, profitable petroleum ventures were beginning to emerge. In 1854 Judge William G. Dryden had acquired what was known as the Tar Well Lot, near what became Westlake Park in Los Angeles. Dryden, believing that the tar overlay a seam of coal, dug a mine shaft on the land. However, at a "shallow depth" the shaft was filled with semiliquid asphalt, and it was abandoned.

A one-third interest in the property was obtained on February 20, 1860, by George H. Gilbert, a former whale-oil refiner. The following month, on March 26, the Los Angeles City Council sold him three other nearby lots, bounded by what are now Hoover and Alvarado streets, Wilshire Boulevard, and Bellevue Avenue. In May of 1860 Gilbert built a 400-gallon retort on the lots to distill liquid asphaltum. By August 29, it was reported that Gilbert had produced 2,350 gallons of distillate, which, after being further refined in San Francisco, was being sold as kerosene at a price of $1.75 per gallon. The next year he moved his refinery to the Rancho Ojai, which had long been used as a local source of crude. Although it burned, Gilbert rebuilt the factory and dug a sixty-two-foot shaft to acquire better crude. When the primitive refinery burned again, he gave up.

The next important development occurred in November, 1863, when Edward Conway and Captain James H. White formed E. Conway & Company. Work began shortly afterward on an eighty-seven-foot well near Canada de la Brea, about six miles east of Richmond. In addition, a primitive refinery was built. By December 12, 1863, Conway and White had invested twenty-five-thousand dollars in the venture. But the project proved unprofitable.

The next year, in 1864, White gained title to 78,000 acres of land near San Buenaventura (or Ventura). Included in White's land was the 48,800-acre Mission San Buenaventura, which he bought for two dollars an acre from the federal government. The acreage included Sierra de Azufre or Sulphur Mountain. White wasted little time in mining a "mass of petroleum," which when refined yielded 38 percent illuminating oil and 48 percent lubricating oil.

In 1866 the Buena Vista Petroleum Company constructed a crude refinery about three miles northwest of McKittrick. The region was well known for its heavy tar-like oil seeps. The tar was hauled to the still and melted over a fire as the workmen fought off the horrible fumes. Many would-be oil prospecters dug shallow wells near asphalt deposits and dipped the tarry substance to the surface. Most of the pits were dug by hand, because it was almost impossible to sink a bit through the asphalt. The deeper the bit penetrated the asphalt, the harder it was to turn. Eventually the bit would stick, and no amount of pulling could free it. The only way to retrieve the bit was to place a heavy weight on the windlass and let the steady pressure slowly pull the bit to the surface.

One of the most active of the early California oil men was Colonel Thomas A. Scott, vice-president of the Pennsylvania Railroad. Scott had first become interested in the petroleum industry in the Pennsylvania oil fields. He later joined J. E. Thomson, head of the Pennsylvania Railroad, in a project to purchase California oil lands. Initially, the would-be oil entrepreneurs sent Professor Benjamin Silliman, Jr., of Yale University, to examine potential petroleum-producing areas of California in the spring of 1864. Later that year J. W. Henderson examined the region around Humboldt County for the two railroad men. He found samples of oil seepage that closely resembled the Pennsylvania crude. Eventually, the Scott group bought nearly 265,000 acres in what became Ventura County and 12,000 acres in Los Angeles and Humboldt counties.

After forming the Philadelphia and California Petroleum Company, the California Petroleum Company, and the Pacific Coast Petroleum Company in 1865, Scott and

Thomson sent Thomas R. Bard to California to manage the effort. When Scott died in 1881, Bard, who handled the dispensation of his business affairs in California, joined part of Scott's old holdings with those of two other Pennsylvanians who had migrated to the California oil fields to form the Union Oil Company of California, which was incorporated in 1890. When Bard arrived in January of 1865, Henderson was sent to look after the Humboldt County property.

The Philadelphia and California Petroleum Company, under Bard's direction, drilled six wells on Rancho Ojai along the northern edge of Sulphur Mountain in Ventura County. Although the first five were dry, the sixth, completed in 1866, flowed at about twenty barrels per day. With no market for the crude, however, the wells were abandoned.

On March 25, 1865, the Union Mattole Oil Company was incorporated, with Thomas Richards, William Ede, and Edward Bosqui, all San Francisco businessmen, as its directors. Soon afterward they began a well on the North Fork of the Mattole River in Humboldt County. On June 7, 1865, the first shipment of "coal oil" from the well arrived in Humboldt for shipment to San Francisco. There it was distilled by the Stanford brothers—Josiah, A. P., and Charles—and sold for one dollar and forty cents per gallon under the brand of Comet Illuminating Oil. Eventually the Stanford brothers began purchasing shares of the Union Mattole Company.

This was the first commercial sale of refined oil from a well drilled in the state. Eventually Standard Oil Company, which had been involved in the California petroleum industry from the beginning, purchased the refinery. Although Standard Oil Company of California was not founded until 1906, its lineage could be traced back to the 1852 activities of F. B. Taylor & Company, founded by Frederick B. Taylor, who operated an oil and camphine firm in San Francisco; the very earliest development of the Pico Canyon area; and the later growth of the Star Oil Works Company.

In the two years following this first commercial sale of California oil, several individuals and companies, including the Union Mattole Oil Company, the Mattole Petroleum Company, and the Oil Creek Petroleum Company, drilled twenty-four wells along the Mattole and Bear rivers and Oil Creek in Humboldt County. Most of the early wells were less than 260 feet deep. However, in 1867 the Stanford brothers drilled a well to 1,003 feet beside the Union Mattole No. 1 on the North Fork of the Mattole River.

By August, 1865, regular shipments of petroleum were leaving Eureka, the nearest seaport, for San Francisco. Transportation facilities were almost nonexistent. The oil had to be carried in small containers on the backs of mules for thirty miles from the wells along the Mattole and Bear rivers to Centerville. There it was transshipped on bad roads to Eureka, where the crude was loaded on steamships.

Many of the wells along the Mattole and Bear rivers were plagued by cave-ins and other problems. In addition, on March 17, 1865, the federal government ordered an end to the disposition of Humboldt County oil lands. This action clouded the lease claims in the region, and the boom died.

Despite the seeming lack of success, many Californians continued the search for crude. Charles Scott developed the region around the Santa Paula oil seeps and opened a small refinery there. The Stanford brothers and others drove tunnels into seeps on the flanks of Sulphur Mountain and caught the crude as it flowed out. Although such primitive methods claimed only a few barrels per day, they were sufficient to supply the Stanford brothers' refinery in San Francisco.

E. Benoist, who operated an oil-refining laboratory in San Francisco, joined with Stephen Bond to drill a well near Buena Vista Lake, in what was then Tulare County but is now Kern County, located along the eastern end of what would later become the McKittrick Field. The effort failed when the bit stuck in the hole. Benoist and Bond did recover some oil from pits in the same area, however. These pits were usually about twenty feet deep, five feet wide, and eight feet long and could produce about three hundred gallons of crude daily. Benoist and Bond operated their pits from February, 1864, until April, 1867. At that time the local market was flooded with crude, and they abandoned their project.

The Point Arena Petroleum Mining Company began operations near the Point Arena wharf in Mendocino County in 1864. It hoped to convert local bituminous rock into coal oil, but failed. So did later wells in the region. An effort by the Bolinas Petroleum Company to drill in Arroyo Hondo in Marin County, to the north of San Francisco, failed in 1865.

Although numerous attempts to tap California's oil reserves were made in the first seven years after Drake's well, no deposit that could compete with the cheap Pennsylvania crude had been found. Part of the problem was the inadequacy of local outlets and the expense of shipping the oil to more lucrative markets. As a result, the California petroleum industry languished from 1866 until the Southern Pacific Railroad penetrated the region in 1876. Railroads made the transportation of crude and refined products in large quantities at low cost feasible, and thus profitable markets became available. So when the Newhall Field was developed in the mid- and late 1870s, a new oil boom began.

In the first thirty-five years of California's petroleum industry, approximately three thousand wells were drilled in the state. These were shallow holes, usually drilled near well-known surface seeps. A few pipelines carried the crude to nearby railroads, but no large trunk lines were built until late in the nineteenth century. No large, important pools were uncovered prior to 1897, and in that year California's output was only about two-million barrels of crude.

Technological innovations stimulated the California petroleum industry in the 1890s. One such breakthrough was the development of the Allen Pumping Unit. A central power unit used to pump from several wells, the Allen Pumping Unit consisted of "a vertical shaft driven by a bevel-gear." "An eccentric, the upper end of the shaft was attached [to] the wires . . . from the various pumps so that the pull of the pumps balanced one another." The revolution of the eccentric varied according to the gravity of the oil, and the wire cables were connected to a reciprocating jack. Use of the Allen Pumping Unit allowed power to be conveyed for more than half a mile in all directions from the central power plant. This in turn allowed a number of wells to be pumped economically at the same time and thereby made marginal wells profitable.

Extensive use of the Allen Pumping Unit and jack-line pumping methods was made in many California oil fields, especially in the Kettleman Hills region, where the shallow formations and the thick crude made such operations extremely profitable. Like the Allen Pumping Unit, jack-line pumping was a simple arrangement; in it, several wells were powered by rods extending outward from a single power source. It was not uncommon for one motor to operate ten wells in the Kettleman North Dome Pool or in the shallow wells of Torrey Canyon.

In addition, in the final two years of the nineteenth century oil took on a new significance nationwide, as increased demand for crude as a fuel—for locomotives and automobiles—spurred additional searches. Eventually California oil men would uncover twenty-one of the American oil fields that proved most prolific in the first half of the twentieth century. However, of those fields uncovered prior to 1900, only McKittrick, Kern River, and Brea Canyon were major discoveries, and they were located in the final thirteen years before the turn of the century.

Left: Juan Rodriguez Cabrillo, a Portuguese sailing under the Spanish flag, gave the first indication of California's rich petroleum deposits when, in 1543, he recorded that the Indians along Santa Barbara Channel used asphaltum to caulk their canoes. *Security Pacific National Bank Collection, Los Angeles Public Library. Right:* British explorer Captain George Vancouver observed a huge oil slick in the Santa Barbara Channel in 1792. Submarine seeps were numerous in the area. *Henry E. Huntington Library and Art Gallery*

Oil seeps such as this one were abundant in the area that was to become California. The seeps provided early entrepreneurs with crude from which illuminant, axle grease, and paving material were made. Early oil prospectors from Pennsylvania were excited by the presence of the seeps but often met with failure in drilling near the seeps, for California's irregular oil-bearing formations were far more difficult to tap than were those in Pennsylvania. *Union Oil Company of California*

Left: General Andreas Pico, who had fought with his Mexican forces in California during the Mexican War, by 1856 was distilling crude from a seep in northern Los Angeles County for use in the San Fernando Mission, thereby becoming a pioneer in the California petroleum industry. *Henry E. Huntington Library and Art Gallery. Right:* Judge William G. Dryden, like many others of the era, believed that the presence of asphaltum signaled the presence of coal. He purchased the Tar Well Lot near the future Westlake Park in Los Angeles in 1854 and dug a mine shaft. When the shaft filled with semiliquid asphalt, he abandoned his effort to find coal. *Henry E. Huntington Library and Art Gallery*

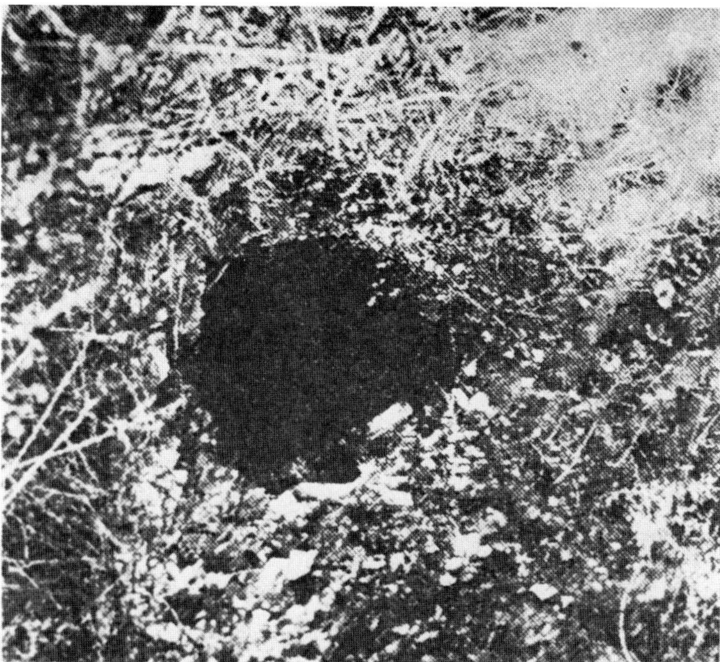

Left: The La Brea tar pits near Los Angeles are the site of one of California's best-known oil seeps. For millions of years animals had fallen victim to the natural trap, which in the nineteenth century began to yield numerous skeletons of prehistoric animals. *Union Oil Company of California. Right:* An opening of one of the early tunnels dug into Sulphur Mountain by early California oil men. Nothing but a crude shaft, the tunnel continued horizontally at a slight angle, so that when the oil stratum was found the liquid would flow by gravity to the mouth of the mine. *PennWell Publishing Company*

The Ojai No. 1 was drilled to the seven-hundred-foot level by A. J. Sallisbury in 1867 on the Rancho Ojai. *PennWell Publishing Company*

Refinery workers pose for the photographer at a Los Angeles plant in 1896. *Standard Oil Company of California*

Among the early refineries in California was this one at Alameda Point in 1880. Note the numerous feeder pipes on the ground and the primitive distilling equipment. *Standard Oil Company of California*

Early geologists exploring California's inland areas for possible petroleum-producing regions often found the going rough and had to depend on mules to carry them through the rugged terrain. *Union Oil Company of California*

The ferryboat *Julia*, which plied the California coast between Port Costa and Vallejo, was one of the first ships on the West Coast to be converted from coal to oil. However, when the ship was destroyed by an explosion, local steamboat inspectors immediately cancelled all other permits for ships to use oil for fuel. *Union Oil Company of California*

A delivery wagon of the Associated Oil Company, used to haul gasoline, which the company boasted provided "more miles to the gallon." The conversion from coal to diesel fuel by the railroads and the development of the automobile provided expanded markets for firms such as this one and stimulated aggressive exploration for California crude. *PennWell Publishing Company*

The first locomotive converted from coal to oil on the West Coast. The initial experiment with using oil instead of coal was carried out by the Union Oil Company of California; when it proved successful, it opened an entire new market for the state's oil industry. *Union Oil Company of California*

The opening of Mulholland Drive in Los Angeles in 1924 signified as much as anything else the arrival of the "Age of the Automobile," which ensured the continued growth of the California oil industry. *Union Oil Company of California*

This horse-drawn tank wagon was owned by the Gilmore Oil Company. Company officials foresaw the coming boom in automobile sales, proclaiming that "some day you will own a horseless carriage" and that "our gasoline will run it." *American Petroleum Institute*

The early oil entrepreneurs in California found oil and gas at extremely shallow depths, although prior to 1897 they uncovered no large pools. Well into the twentieth century, shallow formations were exploited, using a variety of pumping techniques, including this windmill-powered pump of the Burns Oil Company. *PennWell Publishing Company*

To satisfy the need for a means to transport the output of California's early wells, the railroads simply mounted steel tanks on regular flatcars. *PennWell Publishing Company*

Left: Jack lines, a system with a revolving eccentric wheel that alternately pulled and released cables attached to pumps on wells. A single such system could power many individual pumps over a great distance. *Union Oil Company of California. Right:* The pumping end of a jack line. The cable from the revolving wheel was pulled over the vertical wheel, and as it was alternately pulled and released, it raised and lowered the pumper rod in the well. *Union Oil Company of California*

Santa Maria Basin and Its Forerunners

THE Santa Maria Basin contains one giant oil field—Orcutt or Santa Maria—several major producing pools—Lompoc or Purisima, Gato Ridge, Los Alamos, East Cat Canyon, West Cat Canyon, Casmalia, Santa Maria Valley, Arroyo Grande, Los Flores, Doheny Bell—and numerous smaller finds. Its pools stretch in a line from the Purisima Hills, inland from Point Arguello on the coast of Santa Barbara County, northward to just north of Pismo Beach in San Luis Obispo County. However, the exploration that led to the development of the region actually began to the north, along the Southern Coast Ranges.

In San Mateo County, on the southwest side of San Francisco Bay, Bell's Ranch, about fifteen miles below Half Moon Bay, was the scene of tunneling activity into the nearby oil seeps as early as 1865. Other early wells were tried near Bellvale along the shore of Half Moon Bay, about thirty miles south of San Francisco. Eventually four producing areas—Purisima Creek, Purisima Anticline, Tunitas Creek, and La Honda—were developed in a line stretching southward from Miramontes Point to near the junction of San Gregorio and Harrington creeks. It was the most northerly sedimentary basin in California that produced high-gravity oil and gas in commercial quantities. Because of the proximity of San Francisco and a ready market, most of the crude, which was rich in gasoline, was refined locally.

Afterward, oil men continued to work their way southward into Fresno and San Benito counties, where they found the Cantua and Vallecitos pools along the southeast edge of the Diablo Range. The region, about twenty-five miles north of the town of Coalinga, was well-known for its oil seeps. Beginning in the 1880s and continuing as late as 1912, people hand dug numerous wells in the area. Union Oil Company of California and several independents drilled wells in the region, which, varying in depth from seventy to eighteen hundred feet, produced one to five barrels daily—not enough to set off an oil rush.

To the north, Contra Costa County was also the scene of early-day petroleum activity. A well bored in 1862 and one drilled in 1864 both proved unprofitable. Several wells were drilled along Oil Creek on the southeast side of Mount Diablo, but production was never established in commercial quantities. In the twentieth century several dusters were drilled near Altamont, Alameda County.

In 1861, just two years after Drake's discovery in Pennsylvania, workmen cutting logs in Moody's Gulch near Lexington in Santa Clara County cut down a giant redwood that was so large that it gouged a deep hole when it struck the ground. When the tree was removed, the hole seeped full of a dark liquid. Suspecting that the liquid was crude, some of the lumbermen scooped out some and put it in a lamp. When the lamp was lit, it burned brightly.

Several other oil seeps were located nearby, but it was not until April, 1865, that the Santa Clara Petroleum Company drilled a well in the area and the Shaw and Weldon Petroleum Company struck a "vein of finest quality" at thirty feet. That same year the Pacific Petroleum Company began operations near Lexington and by October was shipping crude to San Francisco for transshipment to New York City.

During the late 1870s and 1880s, oil men began drilling along a westerly tributary of Los Gatos Creek on the eastern edge of the Santa Cruz Mountains. Many oil men hoped the discovery, named the Moody Gulch Field, would become the major supplier for San Francisco, sixty miles to the north, which was the big oil market on the Pacific Coast. One of the early developers at Moody Gulch was W. E. Youle, who joined D. G. Scofield in examining the seeps in the canyon. Before securing leases to the region, Youle and Scofield decided to examine the area firsthand. While they were in the field looking for possible drilling sites, R. C. McPherson, in a short-lived partnership with Colonel Boyer, who owned the San Francisco and the Santa Clara oil companies, leased the area. Boyer hired Youle to drill on the property. Of the drilling crew, only Youle had ever worked on a drilling rig before. Furthermore, just getting the proper equipment to the drilling site proved to be a challenge. "I ordered lumber and rig timber sawed at a mill in the Santa Cruz mountains," Youle recalled, and then "I purchased a boiler engine and tools that were stored in San Francisco." The rig had originally been used to drill deep water wells, and an eight-inch drilling bit was the largest available with it. Youle had larger bits custom made. "It was certainly some job to make the forgers understand what oil tool work required," Youle recalled. He purchased "several sizes of pipe from George W. Biggs, an iron merchant of San Francisco," and then "located a 1,500-foot coil of [manila] drilling cable and sandline"—all before drilling could be started.

In the meantime Scofield had persuaded U.S. Senator Charles M. Felton and Lloyd Tevis, president of the Wells Fargo Bank, to examine the Moody Gulch region. Impressed by what they saw while visiting the area, Felton and Tevis provided the financial backing for Scofield to form the Pacific Coast Oil Company, by buying out the California Star Company, the San Francisco Oil Company, the Santa Clara Oil Company, and several other smaller oil concerns. Felton was named president of the new company, but Scofield actually ran the business.

Moody Gulch did not turn out to be a major strike, and only fourteen to sixteen wells were ever completed. Most drilling was stopped by 1912. Total production as of 1921 was about eighty-five-thousand barrels, and after that most of the wells produced only intermittently. In fact, the most important product of the field was probably the Pacific Coast Oil Company, which was to become one of the giants of California's petroleum industry and to help revolutionize the oil business by expanding markets through the use of ocean-going tankers. Early in California's oil history, the company's founders realized that the West Coast offered few profitable markets for crude. In the 1880s the entire population of California was approximately 1,000,000 people, with 250,000 of them living in the San Francisco area. Los Angeles's population in 1882 was a mere 14,000. Obviously some means of getting the state's plentiful crude to other domestic or foreign markets had to be found, and the obvious route was by water.

Primitive tankers had been used to transport California crude to market for years. In 1864, Benjamin Silliman, Jr., president of the Pennsylvania Rock Oil Company, landed in California on his way to check mining properties in Nevada. Examining oil wells around San Buenaventura, he suggested that local oil men utilize vessels designed along the lines of Pacific whalers, which had large tanks to hold whale oil, to transport crude from California ports to other markets. Such makeshift vessels filled the void until June 1, 1887, when Sespe Oil Company ordered a ship with a 3,600-barrel capacity from the Fulton Iron Works at a cost of forty-thousand dollars. Launched on January 10, 1889, the ship was named the *W. L. Hardison* and could carry a deckload of lumber, as well as barrels of oil. Oil tankers were still new, however, and on June 25, 1889, while the *W. L. Hardison* was being loaded with oil at Ventura, an inexperienced first mate lowered a lantern into an oil tank to see how much crude had been loaded. The ship burned to the water's edge.

It was not until 1894 that the Pacific Coast Oil Company and the Union Oil Company of California cooperated in building the "first true tanker on the Pacific Coast." Built with one great oil-storage tank "divided into smaller ones by stiffening bulkheads," the ship was designed to do nothing but carry crude. Christened the *George Loomis*, the vessel left Ventura on its maiden voyage in early January, 1896, to open a new era in the marketing of California's crude.

To the south of the Moody Gulch strike, about seven miles south of Gilroy, oil men were attracted to the natural oil seeps along La Brea Creek on the Sargent Ranch. Drilling began in 1864 and 1865 and after a several-year hiatus, resumed in 1886. By 1904, the Sargent Field, as the area was named, had produced about 20,000 barrels of crude. Sargent's production peaked in 1909, with an annual output of 63,780 barrels. As of January 1, 1941, its estimated cumulative production stood at 600,000 barrels.

By the late nineteenth century most California oil men had concluded that the Southern Coast Ranges south of San Francisco held little crude and had begun to move farther south into the Santa Maria Basin. As early as 1898, oil men began drilling near the natural oil seeps along the Huasna River and its tributaries in southern San Luis Obispo County. By 1942, thirty-one wells had been drilled in what was called the Huasna Area.

The biggest field in the area, however, was the Santa Maria Field, located between the town of Orcutt and San Antonio Creek, just south of the San Luis Obispo–Santa Barbara county line. The area had attracted pioneering California oil men for years. In the winter of 1865, Thomas R. Bard, who would be one of the founders of Union Oil Company of California in 1890, and Charles Fernald, a local attorney, spent a week on horseback examining what became the Lompoc, Cat Canyon, and Orcutt fields.

In 1887 M. P. Nicholson began drilling for water in downtown Central City, which was later renamed Santa Maria. Unsuccessful in finding water, he turned to wildcatting for oil in the region. However, after his death in 1888 (from a kick by a horse), oil men ignored the area until 1902, when the Western Union Oil Company opened the pool to production. Then when the Pinal Oil Company's third well in the pool blew in as a gusher, rapid development of the area started.

The Western Union Oil Company was owned by Union Oil Company, a Delaware-based firm that was the forerunner of the Shell Union Oil Corporation, a part of the worldwide Shell organization. For nearly five years, the field was extensively developed by the Union Oil Company of California, which controlled almost 75 percent of Orcutt's production. However, because of Orcutt's isolation, the field never became a major part of Shell's operations and activity had practically ceased by 1907. A second boom period was triggered when America entered World War I, but by 1922 it too was over and the field again lapsed into neglect.

During its two booms, Santa Maria's productive area had grown to 3,375 acres containing 321 wells, 64 of which were later abandoned. By 1940, the field had produced an estimated 50,000 barrels of oil per acre. In spite of the fact that the two boom periods combined lasted only a decade, Santa Maria's output was great enough that by the close of the first half of the twentieth century it ranked as the seventy-second most productive oil field ever found in America. Total production between its discovery and the close of 1949 was 108,970,000 barrels of oil.

With the discovery at Santa Maria, the Union Oil Company of California dispatched its foremost geologist, William W. Orcutt—who would become vice-president in charge of production for Union and the "dean of petroleum geologists"—to map the surface geology of the entire Santa Maria district. After receiving his report, Union Oil and Lyman Stewart leased nearly seventy thousand acres in the area. However, the leases were short term, and it became a challenge to complete the exploratory wells before the leases expired.

Frank Hill, drilling superintendent on the Hill No. 1, a wildcat in the Purisima Hills near Lompoc, recalled, "We had to race against time because the option was a short one." "We landed in Lompoc the night of July 4, 1902, with our steam cable drilling rig," Hill continued, "by the next day we had located the drilling site, staked out roads, [and] knocked together a bunkhouse. . . . Then we put up tanks and built the derrick. . . . Seventeen days after we hit the field, we spudded in our first well," Hill explained. "We drove ourselves day and night. . . . Shift hours were forgotten in the excitement." Three days before the expiration of the leases, "we struck oil." Trying to keep the discovery a secret, Hill took a train to Los Angeles to report to Stewart. It was Sunday when he arrived and, catching a ride with Frank Garbutt, one of Union Oil's directors, started for Stewart's house. Just as they arrived Stewart stepped out his front door with a Bible under his arm. A deeply religious man, Stewart was known for never doing business on Sunday.

For a minute the three looked at one another, and then Stewart asked what they wanted. "We've got an oil well," was the answer. "What does it look like?" Stewart inquired. "Pretty heavy," Hill answered, showing Stewart a sample. Stewart "stuck his finger in and rubbed the smelly stuff back and forth between a finger and thumb" and then said, "We'll take up the options tomorrow. . . . This is Sunday." Stewart then walked down the street to church.

Lompoc was developed by the Union Oil Company, which eventually drilled thirty-seven wells in the region. In 1903, Frank Hill made oil-field history on one of the wells. While drilling, he discovered that water from higher strata so diminished production that some suggested that the field be closed. Drillers had tried everything to stop the flow of water, including throwing shavings and chopped rope down the hole. They were unsuccessful. To solve the problem, Hill designed a bailer that allowed cement to flow out of its holes. After the bailer was filled with cement, it was lowered to the bottom of the hole, where its cement seeped out and temporarily stopped the flow of water. Encouraged, Hill lowered a packer to the bottom of the well and then pumped cement down the hole. When the cement reached the bottom, it flowed around the outside of the casing and back toward the surface, effectively sealing out the water.

This was one of the first deep-hole cementing jobs in the petroleum industry. Another oil-field improvement developed in the Santa Maria Field was the continuous distillation method of refining oil. The first gasoline extraction plant in California was also constructed in the pool.

Production quickly grew to the point that it became obvious that a town was needed to house the many workers rushing to the strike. Orcutt was called in to lay out the townsite. He selected a grainfield in an "open stretch of country through which the Pacific Coast narrow gauge railroad wound its way from Santa Ynez to San Luis Obispo and Port San Luis." Once the site was staked, "store buildings, warehouses, machine shops, etc., sprang up over night." A siding was built along the railroad, and pipelines were built to carry the crude to the town, where it could be loaded for shipment.

In recognition of his work in developing Union Oil Company's properties throughout California, the new community was named for Orcutt, who was "credited with the discovery of more of California's oil fields than any other individual." In accepting the honor, Orcutt replied that it reminded "him of the practice of naming cheap cigars after cheap actresses." Orcutt went on to become the prime mover in the organization of the Pioneer Petroleum Society of California. Later, in the 1930s, his name was applied to the oil field surrounding the town as well.

Probably the most famous well drilled in the Santa Maria Field was "Old Maud," officially named the Hartnell No. 1. Old Maud was a mistake, as Jack Reed, a member of the well's drilling crew, remembered. "We really caught it after we started drilling," he recalled. "Back in . . . 1904, nobody thought a few feet one way or the other made much difference," Reed explained, so "our rigging crew was supposed to put the enginehouse here and the derrick there." "It was a hot day," he continued, "and when the boiler accidentally fell off the wagon where the derrick should have been, we left it right there and put up the derrick where the enginehouse should have been." "The boss was mad, but not quite mad enough to make us tear down a whole day's work and start over again."

Old Maud was spudded in on June 22, 1904, and on December 2 of that year, "when no one was expecting much of a well," it started "rumbling." "Then with a roar," Reed recalled, "a column of oil and gas shoots up through the rig floor to a height of 150 feet." "Oil begins pouring down the gullies and creek beds," he continued. "We have the biggest producer the world has ever seen," Reed explained; "we can't control it, what with 12,000 barrels of oil pouring out every day. . . . We don't even have tanks or pipelines big enough to handle the flow, so we scrape up a series of earth dams." At the time of its completion, the Hartnell No. 1 was the largest producing well ever drilled in North America.

Accidentally, one of the workers closed a valve. According to Frank Hill, drilling superintendent for Old Maud, "this completely shuts in the well, which is under great gas pressure." Then the oil started "flowing through the formations and for hundreds of yards around every squirrel and gopher hole begins to spout oil," he explained, until "the surrounding fields are full of miniature geysers of oil." Finally, the valve was reopened.

The flow continued for three months. In the first one hundred days of its life, Old Maud yielded one-million barrels of crude. In the three years after its completion, its output was approximately three-million barrels. Old Maud produced for fourteen years before the casing collapsed and sealed the well. Because it was cheaper to drill an offset than to reopen Old Maud, the Hartnell No. 7 was spudded in sixty-five feet away, on the original site where Old Maud should have been drilled. Ironically, the Hartnell No. 7 never produced more than ninety-five barrels per day. As a result of the wartime demand for crude, Old Maud was cleaned out in 1943 and began to flow at a rate of 175 barrels daily.

In May of 1922, the Union Oil Company closed its Lompoc wells. However, six years later, in 1928, it resumed drilling in the field, and in 1929 dry gas and oil injection were started in hopes of increasing production. The next year, 1930, the Vaqueros-Major Oil Company, which later became a part of the O. C. Field Gasoline Corporation, completed its Union Annex No. 1 in the east end of Lompoc. Spurred by this find, which produced six hundred barrels per day, the York Oil Company completed several additional producers in the west end. These wells were eventually acquired by the Alphonso E. Bell Corporation in 1937.

By 1941, fifty-seven wells had been drilled in the Lompoc Field. Thirty-six were producers, thirteen had been abandoned, and eight had proved to be dry holes. Although Lompoc was not known for its high levels of production, one well, the Union Oil Company's Hill No. 4, flowed for twenty years, a record for a California producer as of 1941. Total production as of January 1, 1940, was 7.9 million barrels of crude.

In San Luis Obispo County, about three miles north of the town of Pismo Beach along both sides of Pismo Creek, the Arroyo Grande Field—also called the Tiber, Edna, and Oak Park Pool—was uncovered in 1906. Most of the early development at Arroyo Grande, which was inside the boundaries of the Rancho Corral de Piedra Spanish land grant, was undertaken by the Associated Oil Company. By 1938, the Dolly Adams Oil Company had acquired control of the pool and had begun new exploration in the vicinity. Southeast of the original discovery, the Elberta No. 1 located the Oak Park extension in 1927.

By 1900 oil men were examining portions of western Santa Barbara County, but it was not until 1917 that the Doheny Petroleum Company uncovered the Casmalia Field between Point Sal and San Antonio Creek. The discovery well, the Soladino No. 2, was brought in on January 6, 1917. Initially Casmalia was developed by the Doheny, Pacific, and Associated oil companies; however, Union Oil Company of California drilled on some wells on the eastern edge of the structure. By 1926 most of the early wells were shut down, but in 1930 the O. C. Field Gasoline Corporation reopened ten wells in the western part of the pool to supply its refinery in Casmalia.

Eventually, Casmalia's boundaries were expanded until they were three miles long and a quarter of a mile wide. Half of the twelve hundred producing acres were operated by the O. C. Field Gasoline Corporation. The remaining area was developed by the Richfield Oil Corporation, Associated Oil Company, the Escolle family and their heirs, and the Soladino Land and Cattle Company. By 1941 Casmalia had produced 7.5 million barrels of crude.

Another series of pools, collectively called the Cat Canyon Field, were found in northwestern Santa Barbara County, just to the south of where the Sisquoc and Cuyama rivers join to form the Santa Maria River. Development in the region began with the location of the West Cat Canyon Field, the westernmost of the pools, in 1908 by the Palmer Union Oil Company. After being cleaned, the discovery well, the Palmer No. 1, which later became the Palmer Stendel Oil Corporation's Blochman No. 1, flowed at between six thousand and ten thousand barrels daily.

In 1909 the Palmer company completed its Palmer No. 2, a well flowing at a rate of eighty-five hundred barrels a day, in the West Cat Field. At the same time, the Union Oil Company's Bell No. 5 came in at between five thousand and six thousand barrels every twenty-four hours. Such production quickly attracted oil men to the region, and within a short time the Union Oil Company of California, the Palmer Stendel Oil Corporation, the Gilmore Oil Company, and the Standard Oil Company of California controlled most of West Cat's production.

Previously, in 1904, the Gilmore Oil Company, Ltd., had drilled a hole in East Cat Canyon to a depth of 3,563 feet before abandoning it. Although the earlier effort had been unsuccessful, the Brooks Oil Company, encouraged by the high production of the West Cat Canyon Field, resumed drilling operations in East Cat Canyon in 1909. Brooks completed a 150-barrel-per-day well in the area and triggered additional activity by the Pinal Dome Oil Company, New Pennsylvania Petroleum Company, and West Oil Company. In 1915 the Palmer Union Oil Company at last located a paying horizon. Within a short time the United Consolidated, Union of California, Henderson, Stone-Goodwin, and Santa Maria oil companies were all active in the East Cat Canyon area. During this second period of activity, between 1909 and 1919, a total of twenty-one wells were drilled in the pool, most by the Palmer Union and Brooks companies.

After completing its 1909 well, the Brooks Oil Company connected the field to the Pacific Coast Railroad with a six-inch pipeline. Later, in 1911, the Associated Oil Company built an eight-inch pipeline into the East Cat Canyon Pool. After 1911, most of East Cat's crude was shipped from the Associated Oil Company's marine terminal at Gaviota.

A third wave of activity began at East Cat Canyon in 1924, when the Marland Oil Company tried to tap deeper production with the use of rotary rigs. The deep crude proved to be either too heavy or too low in quality to be commercially viable, and all drilling activity stopped in 1930. Nonetheless, East Cat's estimated cumulative production as of 1937 was 3,750,000 barrels.

At the same time that oil men were first examining East Cat Canyon in 1904, they were looking at the Gato Ridge Area, just to the south. The result was initially the same—no oil. In 1915, though, the Shaw Ranch Oil Company found production at the three-thousand-foot level. Encouraged, the following year the Standard Oil Company drilled another well, which also reported a show of crude. The results were so poor, however, that no additional drilling took place until July 23, 1931, when the Barnsdall and Rio Grande oil companies completed their Tognazzini No. 1, which flowed at 1,098 barrels of crude daily. Drilling in the area peaked in 1937 with nine wells and one dry hole.

The last of the string of pools in the Santa Maria Basin was located in the northwestern corner of Santa Barbara County between the Santa Maria River and the town of Orcutt. Wildcatters began drilling in the area in 1912, but it was July 15, 1934, before the Union Oil Company of California completed its Moretti No. 1, with an initial flow of 50 barrels daily, to open the Santa Maria Valley Field to production. However, the discovery still did not excite local oil men until Union completed its Adams No. 1 on March 31, 1936; this well flowed at 2,376 barrels per day. The Santa Maria Valley Field was eventually expanded to five thousand acres. It was a little more than seven miles long, east and west, and approximately two miles wide. Between April 15, 1937, and June 30, 1938, its output was 3 million barrels of crude.

Much of California's early production was marketed locally as a lubricant or illuminant by door-to-door peddlers like this man, C. H. Driggs, a salesman for the Perfection Oil Can company of Berkeley. *Standard Oil Company of California*

With the discovery of more and more California oil fields, local markets became swamped with petroleum, and new markets were needed badly. One of the easiest means of transporting the crude to other markets was by sea. At first a number of standard sailing ships, such as the *John Ena* shown here, were slightly modified to carry barrels of crude in their holds. Because they were improvised, such vessels proved unsatisfactory; nonetheless, they remained in service for many years. *Standard Oil Company of California*

In 1889 the headquarters of the Pacific Coast Oil Company was located at the corner of Montgomery and California streets in San Francisco. With a population of nearly a quarter of a million, San Francisco was looked upon by most early California oil men as the most important market in the state. Thus when a strike was made at Moody Gulch, just sixty miles to the south, their hopes soared. Unfortunately, the discovery proved to be a disappointment. *Standard Oil Company of California*

The *W. L. Hardison* burning at a pier at Ventura in 1889. Built by the Sespe Oil Company specifically to carry crude, the *W. L. Hardison* held 3,600 barrels of crude in several tanks and cost $40,000. It was set afire when a crewman lowered a lantern into one of its oil tanks to see how much crude was in it. *Union Oil Company of California*

The *George Loomis*, which made its maiden voyage in 1896, was the first "true tanker" built on the West Coast. Learning from the experience of the *W. L. Hardison*, the builders of the *George Loomis* provided a single tank in which to carry crude. Notice that although the George Loomis was steam powered, it also was equipped with sails to proved auxiliary power and to cut the cost of fuel. *Standard Oil Company of California*

As late as 1900, sailing ships were still carrying crude from California to Hawaii. A full-rigged schooner, the *Santa Paula* held 8,200 barrels of oil and was the second tanker purchased by the Union Oil Company. *Union Oil Company of California*

Union Oil Company's *Lyman Stewart* on the rocks at the entrance of San Francisco Bay in 1922. The ship had collided with another ship prior to being driven onto the rocks, where it lay for weeks, pounded by the waves, before finally sinking. *Union Oil Company of California*

The barkentine *Fullerton.* After the sailing ship had given way to the steamship, Union Oil Company of California still utilized the older vessels by filling them with crude and then towing them behind steamships. By such a method twice as much crude could be carried at little increase in cost. The *Fullerton* was safely towed the twenty-five hundred miles from California to Hawaii with sixteen thousand barrels of oil in her hold. *Union Oil Company of California*

Eventually construction of West Coast tankers evolved into a basic design, as shown here in the *Whittier*, which placed the engines and bridge at the rear of the vessel and made the oil tanks an integral part of the hull. This gave the ships more stability in rough seas and greater protection from fire. *Union Oil Company of California*

The *Richlube*, launched in 1926 by the Richfield Oil Company and specifically designed for operation in coastal waters, was powered by a diesel-electric engine but still carried a sail for auxiliary power. Riding low in the water, the *Richlube* obviously was heavily loaded at the time of this photograph. *Atlantic Richfield Company*

Adapted from the standard design of engine and bridge in the rear, the *Richfield*, shown here, has its engine in the stern and its bridge in the bow. The ship, originally named the *Brilliant*, was acquired from Standard Transportation Company by the Richfield Oil Company in 1925. It was capable of carrying twenty-eight thousand barrels of crude and was designed specifically for service along the West Coast. When the ship was loaded, the vessel rested in the water at the painted waterline. When it was empty, the line was clearly above the surface. *Atlantic Richfield Company*

Loading the *Richfield* with crude from the Los Angeles Basin at Long Beach Harbor. The oil was brought to the wharf from nearby refineries or gathering tanks by pipelines. Flexible hoses mounted on wheels, shown in the foreground, were used to transfer the oil from the pipelines to the tanks inside the ship. *Atlantic Richfield Company*

California's oil tankers were prime targets of Japanese submarines after December 7, 1941. The *Montebello* was sunk within two weeks of the attack on Pearl Harbor two hours out of Avila. After hitting the *Montebello* with a torpedo, the Japanese submarine surfaced and finished sinking the ship with its deck guns. *Union Oil Company of California*

The Richfield Oil Company's *Sparrow's Point* was an advanced design for use on the high seas. Although its engines were still located in the stern, the bridge had been moved aft to just slightly forward of the center of the ship. Empty and riding high in the water (in this photograph) the *Sparrow's Point* is being moved by a tug to a loading facility so that its tanks can be filled with crude. *Atlantic Richfield Company*

Thomas R. Bard (*left*) and Lyman Stewart (*right*), who with Wallace Hardison later founded the Union Oil Company of California. Bard initially examined the region between the town of Orcutt and San Antonio Creek in 1865. In 1902, when the Santa Maria Field was uncovered in the area, Stewart leased 70,000 acres in the field. He used much of the profits to finance the early development of the Union Oil Company. *Union Oil Company of California*

The Careaga No. 3, discovery well of the Santa Maria, or Orcutt, Field. The wooden derrick and much of the other original drilling equipment had been removed when this photograph was taken. In its place a huge wooden pumping beam had been installed. Power from an engine in the sheet-iron-covered structure rotated a wheel that raised and lowered the rear of the pumping beam. In turn the suction rod, connected to the front of the beam, rose and fell in the hole as it pulled the crude to the surface. *American Petroleum Institute*

Left: To service the oil tankers carrying Santa Maria crude throughout the world, a large tank farm was constructed at San Luis Obispo. One of California's most spectacular and costly fires was caused by lightning at the tank farm on April 7, 1926. In this night view of the inferno, the tank in the background had just exploded, sending a huge fireball into the air. *Union Oil Company of California.* *Right:* One of the stranger-looking drilling rigs in the California oil fields, this affair towered above the well head. it was powered by a steam boiler, and the walking beam was raised and lowered by the huge shaft to the right of the twin beams that support the walking beam. The cuttings lifted from the hole have simply been piled by the drill floor. Apparently the rig was designed to be reused and could be lowered to be hauled to another drill site. *Pacific Oil World*

Tool dressers Fred Erdman (*left*) and Len Morrow, cleaning boiler flues and fireboxes at a stationary boiler. Normally tool dressers assisted the driller in the operation of the rigs and were responsible for keeping the cable tools in good working order. Sheet-iron walls have been constructed around the boilers to protect them from sudden gusts of wind. *Security Pacific National Bank Collection, Los Angeles Public Library*

Left: Frank Hill, known as "Handy Andy" to his co-workers, the drilling superintendent of the Hill No. 1, the discovery well of the Lompoc Field. To overcome the underground water seepage that threatened to ruin many of the wells in the field, Hill invented the process of using cement to seal the well hole. Eventually the process was adopted industry-wide and became a common practice. In this photograph, Hill is standing beside the remains of the Lakeview No. 1. *Union Oil Company of California. Right:* The most famous well in the Santa Maria Field—"Old Maud," officially named the Hartnell No. 1. It was completed December 2, 1904, producing 12,000 barrels per day. Within three years, Old Maud yielded 3,000,000 barrels of crude. At the time of its completion the well was the largest oil producer in North America. *American Petroleum Institute*

A primitive cable tool rig sinking a well in a barnyard at Lompoc. The triangular derrick supported the tools in the hole, and the steam boiler in the background supplied the power. Judging from the small derrick, this is probably a very shallow hole. *Security Pacific National Bank Collection*

These men are "swinging spider" on a cable tool rig. In this process the casing is kept moving while hole is still being made. *Security Pacific National Bank Collection, Los Angeles Public Library*

Dressing a kitt, or sharpening a cable-tool drill bit, around 1907. The men in the photograph are restoring the proper angle to the cutting edge of a bit with sledgehammers. It was the responsibility of the tool dressers on each cable tool rig to ensure that the drill bits were in top shape. *Security Pacific National Bank Collection, Los Angeles Public Library*

The interior of a typical cable tool rig operating in the Santa Maria Basin in 1909. F. W. Jenning, drilling superintendent; Earl Cay (?); Jim Collins; Bert Wheat; and Allison (?). The men are getting ready to lower the bit back into the hole after bailing out the cuttings. *Pacific Oil World*

By 1910 enough production had been developed around San Francisco to supply not only that city's need, but also that of neighboring communities. This tank truck, belonging to the Standard Oil Company of California, was the first of its kind in Oakland. *Standard Oil Company of California*

The Richfield Oil Company, which marketed Richlube Motor Oil and which became a part of Atlantic Richfield Company, was one of the major developers of the Santa Maria Basin. Although this photograph was probably taken in the early 1930s, the truck was still equipped with solid tires. *Atlantic Richfield Company*

The 8,000-barrel Dubbs cracking unit refinery of the Wilshire Oil Company at Norwalk, California. Jesse Dubbs and his son Carbon Petroleum Dubbs designed a system to demulsify the crude from the Santa Maria Field, which was almost impossible to distill by any other method. Under the Dubbs system, the oil was heated under pressure generated by the vapors given off by the crude as its temperature increased. *PennWell Publishing Company*

The Santa Barbara–Ventura Basin

TO the north and west of Los Angeles is the Santa Barbara–Ventura Basin. This rich oil- and gas-producing region stretches from the Pacific shore near Point Conception in Santa Barbara County southward to Ventura and then inland on a line along the Santa Clara River and its tributaries in Ventura County to a point just west of the town of Newhall in Los Angeles County. As early as the 1860s, oil men were developing the region. The first activity took place in what became known as the Santa Paula Field, a series of small producing areas—Aliso Canyon, Wheeler Canyon, Salt Marsh Canyon, Adams Canyon, and Santa Paula Canyon or Arroyo Mupu—stretching along the south side of Sulphur Mountain, about five miles northwest of the town of Santa Paula in Ventura County.

Sulphur Mountain's sides were too precipitous to allow the construction of derricks and engine houses, so the only way to reach the crude was by conventional mining methods. In 1866 Leland Stanford completed an eighty-foot-long tunnel into the south slope of Sulphur Mountain. Oil flowed by gravity out of the excavation into pits. Forty-five such tunnels were dug. Some were as long as sixteen hundred feet and produced as much as sixty barrels of oil daily. Much of the early crude was refined into lubricating oil for the Central Pacific Railroad.

Periodically along the tunnel shafts, as in other mining operations, upward ventilating shafts were dug to provide adequate air for the miners. To light the interior of the main shafts large mirrors were placed outside the entrance and aligned with the sun to reflect sunlight into the shaft so the miners could work. As the earth rotated, the mirrors also were moved to keep in line with the sun's rays.

The Hardison & Stewart Oil Company, owned by Wallace Hardison and Lyman Stewart and a forerunner of Union Oil Company of California, was one of the early tunnel developers at Sulphur Mountain. One of the firm's tunnels into the mountainside was the Adams Tunnel No. 4, more commonly called the Boarding House Tunnel. By April 4, 1890, it had penetrated 950 feet into Sulphur Mountain under the direction of Harvey Hardison, the brother of Wallace Hardison. Early that morning an explosion occurred deep within the tunnel. Fortunately, no one was killed in the initial explosion, and Wallace Hardison, who had rushed to the mine site, ordered everyone to stay clear of the shaft. Afterward, he left to take four injured workers for medical care. While he was gone, Harvey Hardison, ignoring his brother's instructions, led an inspection party of three other men into the shaft to examine the damage. They were seven hundred feet into the tunnel when another explosion occurred. Harvey and two of the other men were killed. Because of such dangers, oil men eventually replaced the tunnels with wells, some of the earliest of which were drilled by the Hardison & Stewart Oil Company. By 1939, approximately 110 wells had been completed in the Santa Paula Field, which had a cumulative production of 1.2 million barrels as of that date.

Another early discovery was made along Sespe and Piru creeks, to the north of the Santa Clara Valley, in Ventura County. The Union Oil Company opened the Sespe Field in 1887, when it completed its Tar Creek No. 1. In November of 1937, the Colima Oil Company resumed exploration of the Sespe Field, and Colima's second well in the pool flowed at five hundred barrels daily. Continued drilling in the Sespe area expanded the field to include the Tar Creek, Fourforks, Foot-of-the-Hill, Little Sespe Creek, Lime Canyon, Modelo Canyon, and Hopper Canyon pools. Though the heaviness of Sespe's oil hindered development somewhat, its cumulative production stood at 3,098,000 barrels of crude in 1940, and the pool still contained seventeen flowing wells.

At about the same time as Sespe's original discovery, California oil men were starting to drill in what became the Greater Newhall Field. Actually a series of pools—Pico Canyon, DeWitt Canyon, Towsley Canyon, Wiley Canyon, Rice Canyon, East Canyon, Tunnel Area, Elsmere Canyon, Whitney Canyon, and Placerita Canyon—the Greater Newhall Field stretches southeast from the Pico Anticline, near the Ventura–Los Angeles county line, to just south and east of the town of Newhall. The Pacific Coast Oil Company—forerunner of Standard Oil of California—was the prime developer of the field, and it was at the town of Newhall that Standard built California's first gasoline plant in 1916.

The Newhall area was an extremely hostile environment. The climate was subhumid: some areas near the coast, in what was called a "fog desert," received fourteen inches of rainfall during the winter months. The isolation of the region was so pronounced that grizzly bears still roamed the countryside when oil men first penetrated the Sespe and Santa Paula canyons.

Oil seeps first were uncovered in the region during the gold rush of 1849. In 1855 Andreas Pico and his nephew, Romulo Pico, dipped crude from hand-dug pits in what became known as Pico Canyon. A decade later, in 1865, Ramón Perea, a Mexican hunter, trailed a deer into Pico Canyon and located a natural oil seep. He took a sample of the crude to the inhabitants of the Franciscan mission settlement in San Fernando. Later, J. L. Del Valle acquired the sample and showed it to a Dr. Gelsich, who promptly staked a placer claim in the vicinity. In 1869, a spring-pole drilling rig drilled a 140-foot hole in the area, which flowed at between seventy and seventy-five barrels daily.

In 1875 C. C. Mentry, and later the California Star Oil Works Company, started a four-well drilling program in the Pico Canyon region. Although the first three proved disappointing, the fourth well, the California Star Oil Works Company No. 4, initially flowed at one hundred fifty barrels daily. Although its output eventually dropped to thirty barrels per day, it produced for more than half a century.

California's first pipeline was built from Pico Canyon to the town of Newhall in 1879. Later the two-inch line was extended to Ventura on the Pacific coast. In the winter months, though, Newhall's developers had to build fires along the line to keep the thickened crude flowing. Leaks were also a problem. Often, when a leak appeared, angry farmers would rip up the pipe because the crude was ruining their crops.

Some of the pioneer oil men who developed Sespe and Newhall faced the question of obtaining clear title to their early drilling sites. Many of the oil seeps were located on public domain. Under the California Possessory Act of 1852, any individual could claim 160 acres, provided that he spent at least two hundred dollars improving the property—but all the claims had to be "clearly bounded." Of course when hundreds of men rushed to stake claims there were overlaps and confusion. As a result, few individuals were willing to risk thousands of dollars to develop a claim that could be challenged. To bring order to the chaos, early California oil men turned to the organization of mining camps, which, after 1864, the federal government recognized as legal governmental bodies. As early as March, 1865, the Los Angeles Asphaltum and Petroleum District was organized in Los Angeles County. Four months later, the San Fernando Petroleum Mining District was formed in the region around Pico Canyon. In July of 1876, the Sespe Petroleum Mining District was organized.

Generally the early districts adopted rules that allowed a single claim to every miner, except the discoverer of a deposit, who usually was entitled to a double claim. To maintain a claim, the claimant had to perform at least one day's work on the claim each month until his labor equaled the value of one hundred dollars. In addition, the establishment of joint claims was allowed so that the cost of development could be shared. Enforcement of the rules was delegated to a mining recorder elected by the oil men.

As the method of recording claims was systematized, the development at Pico Canyon of the other previously uncovered oil deposits was renewed. In addition, other nearby regions—DeWitt, Wiley, and Elsmere canyons—were probed in the 1880s. In the following decade three others were opened to production: Towsley Canyon in 1890 and Whitney and Placerita canyons in 1893.

By 1902 most pools in the Greater Newhall Field had been thoroughly drilled. As late as 1916, however, some drilling activity was still taking place in Pico Canyon. In 1940 Standard Oil of California shut down its operation in Pico and Towsley canyons. Most of the other canyons in the Newhall area suffered the same fate.

Following the Greater Newhall strike, the Union Oil Company uncovered several nearby small pools. In 1891, the company began examining the region along the Santa Clara River in northern Ventura County. Although Union's Robertson No. 1 found oil, in what became the Bardsdale Area of the Bardsdale Field, the production was not sufficient to make the well profitable. By 1900 most activity had ceased. In 1915, though, Union drilled the Robertson No. 15 searching for deeper horizons which it finally tapped in 1935. By 1936, fifty-five producing wells were operating in the field. Their cumulative production was 1,363,400 barrels of crude.

In 1911, the Shiells Canyon Area of the Bardsdale Field was opened by the Montebello Oil Company. The discovery well flowed at about twenty barrels daily from 581 feet. Although most wells in the pool were completed by 1918, some drilling continued until 1926. As with the Bardsdale Area, deeper production was located. By 1938 the Shiells Canyon Area contained 145 shallow and 14 deep wells. Some of the shallow wells flowed at as much as one hundred barrels per day, but the greatest output came from the deep zones, which produced some seven-hundred-barrel-per-day wells. Total production of the area stood at seven million barrels by 1936.

Shortly after the initial strike at Bardsdale, the Union Oil Company of California uncovered the Conejo Field, in 1892, when it completed the Calleguas No. 1, about two miles south of Camarillo at the foot of Conejo Grade. The oil was shallow, at between sixty and eighty feet, and the find unique in that the production came from volcanic rocks. Conejo proved to be a small producer, however, and by 1940 contained only fifty-three wells, with an estimated cumulative production of twenty-five thousand barrels.

Local residents had long been aware of natural oil seeps in what was originally called Oil Canyon, located in the foothills of the Simi Valley about two miles north of the town of Santa Susana in Ventura County. Then, in the period between 1900 and 1902, the Simi Oil Field was uncovered. Drilled beside a natural seep in Oil Canyon by the Simi Oil Company, the discovery well was only a small producer, but it encouraged other activity. In 1912 the Petrol Company drilled a well in nearby Tapo Canyon that produced a fair amount of crude. The largest producer in the Simi–Tapo Canyon Field was completed in Tapo Canyon in 1913 by the Santa Susana Syndicate. It initially flowed at two hundred and fifty barrels daily from 718 feet. By 1922, fifty wells had been drilled in the field.

Although these early discoveries were small, they led to the location of one of America's greatest oil finds, the Ventura Avenue Field, just to the north of the town of Ventura in Ventura County. Natural gas seeps were common knowledge in the vicinity. One area man, Ralph Lloyd, recalled that while he was growing up on the family ranch in the 1890s, the region was struck by a violent storm. Fearing for some of their cattle, his father left home on horseback to look after the animals. As he was riding across the range, a bolt of lightning suddenly started a brush fire. In attempting to escape the flames, the elder Lloyd fled to a piece of ground that was not covered with vegetation. Unfortunately the lack of growth was the result of a natural gas seep, and he barely escaped with his life when the fumes were ignited by the brush fire; his horse was burned to death.

In 1898, after completing his studies as a geologist, Ralph Lloyd returned to the family ranch to map much of the Ventura Avenue area. Later he convinced Joseph

Dabney to join him in leasing what Lloyd believed was "all the possible oil-bearing land" that was available. To develop their property, they formed the State Consolidated Oil Company.

The first commercial development of the Ventura Avenue Field took place in 1903, when seven shallow gas wells were completed by the Ventura County Power Company, which became the Southern California Edison Company. Twelve years elapsed, however, before additional drilling took place. In May of the following year, 1916, the State Consolidated Oil Company completed its Lloyd No. 1 at a depth of 2,555 feet. The well produced a small amount of crude, mixed with a large amount of natural gas and salt water.

It was this presence of natural gas, usually under tremendous pressure, that would plague the oil men attempting to develop the Ventura Avenue Field. Geologists did not learn until some years later that the region was composed of "faults, cross-faults, dying out of faults" and structures that would suddenly disappear and then reappear a few hundred feet deeper. Drillers had never before encountered such a confusing set of underground formations. "Drilling would proceed only a few feet before a gas pocket of shallow depths would be encountered." If the drilling crew was fortunate, "the gas would blow out only the string of drilling tools; if they weren't so lucky the tools, casing, rig, and all, would go vaulting into the air, a geyser of water and gas would blow for a few days, and then patiently rebuilding the rig, they would start all over again."

At first, between 1915 and 1920, most Ventura Avenue oil men attempted to drill with cable tool rigs, because they believed that the high gas pressure made the use of rotary tools impossible. They enjoyed little success. As Ralph Lloyd recalled, "We had nothing to show but craters blown out of the earth, . . . strings of wrecked casing, . . . and a few puddles of oily salt water."

To overcome the problem, Shell Oil Company, which developed much of the area, designed a "circulating cable tool" rig to master the gas pressure by holding down the gas with mud. However, the process was very time-consuming. In addition, much of the mud was lost through underground fissures. In one instance, a crew suddenly discovered a stream of mud pouring from the ground a short distance from their rig. The more mud they poured into the well, the faster the stream flowed. Because of such occurrences, the experiment proved unprofitable. In four years, Shell spent approximately $4.5 million to complete a well that initially flowed at a hundred and fifty barrels per day.

Eventually much of Lloyd's acreage was developed, in partnership with Shell. Most of the land was contained in four large leases: the Taylor, the Gosnell, the Harman, and the McGonigle. The McGonigle lease was the most difficult to drill. A tall ridge stood between the potential well sites and the nearest road. Building a roadway over the ridge was too expensive, so an aerial tramway was constructed to haul supplies to the property. Because a steam boiler would be too heavy to haul on the tramway, the McGonigle lease was developed by an electric-powered drilling rig—one of the first uses of electricity for drilling in California.

In 1925 the huge natural gas production of the deeper formations of the Ventura Avenue Field was tapped. Because oil men could not cope with the pressure, much of the pool's early natural gas output was wasted. When the loss reached an estimated 40 to 60 percent in 1929, the California Oil and Gas Supervisor issued orders to limit the waste.

Beginning on December 1, 1929, the District Oil and Gas Commission, established by the Gas Conservation Act passed earlier that year, required that natural gas waste not exceed 10 percent of production. At the same time, it ordered the closing of sixty-four wells with high gas-oil ratios. These actions, coupled with the 1930 Associated Oil Company's introduction into the region of rotary rigs, which used heavy mineral muds to hold the gas pressure in check, eliminated much of the natural gas waste. By 1940, it was estimated that less than 1 percent of Ventura Avenue's natural gas production was being lost.

By January of 1941, the Ventura Avenue Field covered two thousand acres and was approximately five miles long and one mile wide at its extremes. On June 30, 1941, the field boasted 449 wells. Maximum production was reached in 1929 at 21 million barrels of crude. After 1932, the output of the pool was curtailed and dropped to less than 13 million barrels. Total production as of September, 1941, was 231.5 million barrels.

Ventura Avenue's production was so great that at the end of the first half of the twentieth century, the pool ranked twelfth among the nation's oil fields in total production. Its cumulative output at that time stood at 376,787,000 barrels of oil. Even more astounding was the length of time that Ventura Avenue's production remained high. As late as 1949 its output of 21,133,000 barrels of crude was enough to make Ventura Avenue America's fifth most productive field in that year.

Following the discovery of Ventura Avenue several smaller fields were opened. The South Mountain Field, along the north side of South Mountain, lay on a "sheer, oak-clothed scarp." Located about two miles east of Santa Paula in Ventura County, the South Mountain Field was opened by the Oak Ridge Oil Company in April, 1916. By 1930 most drilling in the South Mountain Field was finished. South Mountain's cumulative production as of 1936 stood at 19.5 million barrels of crude.

Oil men began moving northward into Santa Barbara County as early as the spring of 1921, when E. J. Miley contracted with H. R. Johnson and Frederick P. Vickery for a geological examination of the coastline to the west of Santa Barbara near Goleta Point. Once at work, the two geologists quickly uncovered an anticline, which offered the potential of profitable petroleum production. Six years passed, however, before the Miley Oil Company completed its Goleta No. 1 on the Tecolote Ranch, about five miles west of the town of Goleta, on February 25, 1927. When the Goleta No. 2 was completed and found to produce four hundred fifty barrels per day, the rush was on.

The Goleta Field became known for the fine quality of its oil. Most of the crude tested 44 degrees Baumé with a gasoline content of 70.8 percent and a kerosene content of 2.5 percent. Some wells in the field proved to be excellent producers, with the Santa

Barbara Oil Company's Hollister No. 2 flowing at 1,040 barrels daily. Shortly after the field's discovery, however, Goleta's wells began to produce water, and by February, 1928, most of the holes had been abandoned.

Both Ventura and Santa Barbara counties faced the Pacific Ocean. It was only a natural conclusion that the oil discoveries along the coast extended offshore. This theory was supported by the presence of oil slicks in the Santa Barbara Channel. The problem was how to tap the underwater deposits. California led the way in pioneering offshore drilling. Primitive efforts were being made in the mid-1880s. In 1894, H. L. Williams uncovered the Summerland Field along the coastline just outside the town of Summerland, which he founded, near Santa Barbara. It was obvious to Summerland's developers that the oil play extended beneath the nearby water, and to reach the crude they constructed piers from the shore to potential offshore well sites. The experiment proved so successful that within a short time fourteen piers stretched into the water.

One pier, constructed by the Southern Pacific Railroad, reached 1,230 feet into the ocean. The wells farthest from shore were in fifteen to twenty-five feet of water at high tide. An interconnecting system of three wharfs joined the piers at Summerland together. The wharfs were constructed of eight-inch by eight-inch cross beams and stringers, with a narrow walk, which varied from one to five feet in width, providing access to the wells. By 1903, 198 producing wells were pumping in the pool.

Offshore drilling was adopted in other California fields along the Pacific Coast. Both Ricon and Elwood saw extensive pier drilling. At Elwood one pier reached 2,300 feet seaward. The greatest stimulus to California's offshore operations occurred in 1921, when the state legislature enacted the Tidelands Leasing Act. The bill authorized the surveyor general of the state to issue prospecting permits to offshore land controlled by the state. Within a short time, several offshore leases were opened off the coast of Orange, Ventura, and Santa Barbara counties. In spite of the lapse of nearly three decades, little new technology had been developed since the earlier attempt at Summerland.

The next innovation in offshore drilling came in 1927 at Huntington Beach. Here oil men spudded in a well onshore but slanted the hole offshore. Whipstocking, as it was called, quickly became a common practice. Also during the late 1920s several oil companies constructed artificial islands from which they drilled. Later, as a money-saving innovation during the Great Depression, some oil men opted not to construct an entire pier but to build just the portion that was needed to house a well. Called "steel islands," they were really drilling platforms.

Offshore drilling from a platform actually started in California after a confrontation between Barnsdall Oil Company and David Faries. The episode began when Barnsdall completed a whipstocked well about seven miles north of Santa Barbara but neglected to file a mineral claim within a year, as California law required. Faries then filed the necessary papers, acquired legal title, and leased the property to the Honolulu Oil Company.

Irate over being outsmarted, René Broomfield, Barnsdall's general manager, refused to allow Honolulu's employees access to the property through adjoining Barnsdall leases. Because legal action to force access would take a long time, Honolulu's head, H. M. Van Clief, looked for another alternative. Acquiring a pile-driving raft and ship, Van Clief had an offshore platform constructed on the lease. The lease, though, was outside the breakers, and it was impossible to sink the piles needed to support the platform. Under a new state law, anything beyond the established high-tide line was out of the control of leaseholders, so Van Clief had the pile-driving raft anchored beyond the high-tide line and ordered a pier started outward from it. Since the raft could be reached across the tidal flat only at low tide, as soon as the tide went out workers hurriedly hauled material to the pier and then rushed back to safety ahead of the incoming tide. Because there was no way to relieve the drilling crew during high tide, working hours also were coordinated to the rise and fall of the tide. Faced with the fact that the well was being drilled in spite of his opposition, Broomfield later relented and opened Barnsdall's property to Honolulu's employees, and the drilling progressed on a more regular schedule.

As offshore drilling technology advanced, so did the interest in drilling for offshore oil deposits. In the early 1930s, the Texas Company acquired the rights to a revolutionary submersible drilling barge developed by Louis Giliasso that greatly increased shallow offshore exploration. It allowed drilling equipment to be towed to the well site and anchored in place. Once the well was completed, the equipment was refloated and towed to the next site.

California's offshore development received a serious setback in 1938, though, with the passage of the State Lands Act, which limited offshore activity to tidelands that already were being developed or that could be tapped by wells onshore. This effectively ended for the time being additional offshore exploration in California.

Most offshore wells continued to be drilled from piers, platforms, or whipstocked wells, since other methods proved too costly. This technological limitation, coupled with the outbreak of World War II and accompanying shortages of material and supplies, postponed offshore development. Not until 1947 would the Kerr-McGee Corporation propose a revolutionary idea: sink an offshore well out of sight of land. Utilizing a specialized war-surplus ship, Kerr-McGee pioneered the development of center-well drilling ships, which could sail, or be towed if their engines had been removed, to an offshore drilling site. Once anchored in place, they could sink a well and then move to another location. Kerr-McGee's first venture was off Louisiana, but the practice quickly spread to California.

During the following decades, entire fleets of various offshore equipment were developed, including self-propelled drill ships, drilling tenders, floating platforms, jack-up platforms, and others. As a result, when the Cunningham Tidelands Act of 1957 opened new areas for drilling activity, California's offshore oil industry could come of age. In November, 1958, Summerland Offshore became the first field discovered solely by offshore technology. Within a short time numerous others, including Calienta Offshore, Conception Offshore, and Gaviota Offshore, were located. By January, 1982, California's offshore production had reached 5,634,000 barrels per month, a quantity sufficient to rank the state second nationally in offshore production.

In spite of these ventures and early success at Summerland, however, most early-day California oil men believed that the coast of the Santa Barbara Channel held little crude. As a result, they ignored the area between the town of Ventura and the Santa Barbara County line. Only in 1921, after the Ventura Avenue Field moved westward, did they began to examine the region carefully.

In 1923 and 1924, the Associated Oil Company drilled its Taylor No. 1-A in Canada del Diablo in what eventually became the Rincon Field. Although the well was pushed to 5,215 feet, the oil men stopped 500 feet short of highly productive horizons and abandoned the hole. It was not until the Pan American Petroleum Company completed its Fee No. 3 on December 24, 1927, that the Rincon Field was opened to production.

As development of the Rincon area continued, the Chanslor-Canfield Midway Oil Company's Hobson No. A-2 became the world's deepest well on May 8, 1931, when it reached the 9,702-foot level. On May 16, it became the first well in the world ever to penetrate 10,000 feet. After being plugged back to 7,825 feet and completed on September 14, 1931, it produced well—more than 270,000 barrels of crude as of April, 1942. Once oil men were convinced that the coast of the Santa Barbara Channel contained oil, several other discoveries rapidly took place. About ten miles west of Santa Barbara, the Barnsdall and Rio Grande oil companies uncovered the Elwood Field on July 26, 1928. Peak production was obtained by 1930, when the field accounted for 6 percent of all the oil produced in California.

Gas seeps had long been known along the mouth of Goleta Slough, near Goleta Point north of Santa Barbara. Then on August 5, 1929, the More No. 1 of the General Petroleum Corporation of California opened the La Goleta Gas Field to production, flowing at 60 million cubic feet from a depth of 4,533 feet. By 1937, the La Goleta Gas Field was expanded to two hundred fifty acres. At the end of that year, total production amounted to 14 trillion cubic feet.

The Western Gulf Oil Company began exploring the Gaviota-Conception region, at the northern and western end of the Santa Barbara Channel, in 1929 and struck a huge pocket of natural gas. That same year the General Petroleum Company uncovered the Capitan Field, a minor find about twenty-four miles west of Santa Barbara along Highway 101. Years later, on September 8, 1940, R. E. Havenstrite completed the Lincoln No. 1 about forty miles north of Los Angeles to the Del Valle Field. The discovery well initially flowed at 400 barrels daily, but after extensive reworking it produced 875 barrels of oil and 280,000 cubic feet of natural gas a day.

Del Valle marked the end of the long string of discoveries in the Santa Barbara–Ventura Basin. For nearly forty years the area had been the scene of some of California's most frenzied oil activity. Moreover, it had hosted the pioneering of America's offshore oil development—an advance in technology that would have worldwide importance.

Left: Wallace L. Hardison, who would become the first treasurer of Union Oil Company, which he helped to found, was one of the leaders in tunneling for oil into the side of Sulphur Mountain in Ventura County. Crude flowed by gravity out of the tunnels into pits. *Union Oil Company of California. Right:* Workers at the Hardison & Stewart Oil Company refinery at Santa Paula in 1887. *Union Oil Company of California*

The Hardison & Stewart refinery at Santa Paula. Constructed in 1887, the refinery had a capacity of approximately fourteen thousand barrels its first year. It produced asphaltum, greases, lubricants, and illuminating oils. It was destroyed by fire in 1896 and was never rebuilt. *Union Oil Company of California*

Men at work in a tool shop owned by Union Oil Company at Santa Paula near the turn of the century.
Union Oil Company of California

Union Oil Company wells near Tar Creek, some of the company's best producers during the 1880s and 1890s, were actually relatively insignificant compared with wells yet to be discovered in the region.
Union Oil Company of California

Left: The corner, second-floor office of this building in Santa Paula was where the incorporation papers for the Union Oil Company of California were signed in 1890. *Union Oil Company of California.* *Right:* A typical well at Tar Creek prior to 1900. The steam boiler in the foreground provided power to raise the heavy drill so it could be dropped repeatedly into the hole, thus pounding its way through the earth. *Union Oil Company of California*

Drillers David Swartz (*left*) and Hall Proudfoot in the derrick house of a Hardison & Stewart well at Tar Creek in 1883. Note the large "bull wheel" and heavy cable used to raise and drop the drill bit when the rig was "making hole." *Union Oil Company of California*

Union Oil Company wells in Adams Canyon in 1890. The crude in this pool, so sour that the smell could be detected for miles, was used primarily for coating streets. *Union Oil Company of California*

Wells of the Union Oil Company of California in the Torrey Canyon Field in 1897. The company's entire production that year was approximately 125,000 barrels, and 56,000 barrels of it came from these wells. *Union Oil Company of California*

Fifteen wells can be seen in this view of Union Oil Company operations in the Torrey Canyon Field in 1900. Steam blasts are billowing from several boilers. Note the supply of casing pipe in the foreground. *Pacific Oil World*

The individual wells are being powered by lines running from a single eccentric wheel located in the sheet-iron building. To keep roaming cattle out of the well sites, each well has been fenced. *Union Oil Company of California.*

A view of the Star Oil Works Pico No. 4, which opened production in Pico Canyon and which generally is considered California's "first commercial oil well." Completed at three hundred feet on September 26, 1876, this well prompted the formation of the Pacific Coast Oil Company, a forerunner of the Standard Oil Company of California. *Henry E. Huntington Library and Art Gallery*

Another view of the California Star Oil Works in the rugged Pico Canyon. California's first pipeline was built to transport the production from these wells to the nearby town of Newhall. *Standard Oil Company of California*

A young worker with wrench in hand poses among the machinery of the Star Oil Works in Pico Canyon. *Standard Oil Company of California*

A view of the machine shop of the Star Oil Works in Pico Canyon. *Standard Oil Company of California*

Headquarters of the Pacific Coast Oil Company, which eventually became the Standard Oil Company of California, in Newhall. Because of the nearby strike, Newhall had become quite a boom town. Note the clapboard buildings and false fronts. On the right is a wheel-mounted diner, which could be pulled from one boom town to another and opened for business simply by raising the shutter covering the front. *Standard Oil Company of California*

The Pacific Coast Oil Company's Newhall refinery as it appeared in 1890. The facility was constructed in 1876 to process crude from Pico Canyon. The refinery had a capacity of 22,000 barrels per year and produced mostly kerosene and grease. *Standard Oil Company of California*

Wiley Canyon wells, drilled in the late 1870s, supplemented the Pico Canyon production. Note how the rig builders had to construct the base of the derrick to compensate for the extreme steepness of the slope. *American Petroleum Institute*

Hardison & Stewart's Robertson Well No. 1 at Bardsdale as it appeared in 1890. This was the firm's last well before it merged with other interests to become Union Oil Company of California. The steam boiler can be seen in the extreme left of the picture. The "walking beam," extending from the derrick, was powered by the boiler in an up-and-down motion, which in turn was used by the driller to raise and drop the drill bit. *Union Oil Company of California*

Shell Oil Company's automatic pumping station in the Simi Field. The automatic pumps are housed to the left of the water tower, and the overseer's home is to the right. *PennWell Publishing Company*

The State Consolidated Oil Company's Barnard No. 1, drilled in 1916. Note the stacks of casing pipe near the well, the second drilling site being prepared to the right, and the citrus grove in the background. *Pacific Oil World*

Shell Oil Company officials dropping the old wooden derrick on Gaswell No. 1, which had been completed in 1919. The Gaswell Lease was aptly named, for the Ventura Avenue field featured natural gas, which erupted under enormous pressure. Wells commonly went "wild," and many of them created huge craters. *American Petroleum Institute*

The Ventura Avenue Field as it appeared in 1929. Note the storage tanks on the ridge in the background. *Atlantic Richfield Company*

A portion of the Ventura Avenue field in 1935. By this time, the field was recognized as one of the nation's most productive oil and gas fields. Despite the extensive development, a number of citrus groves can be seen in the picture. The Pacific Ocean is in the background. *Standard Oil Company of California*

Drilling rigs in the South Mountain Field usually were perched in precarious positions on the tops of ridges, while pipelines were laid wherever the terrain allowed. *PennWell Publishing Company*

Left: The rugged terrain at South Mountain forced the Union Oil Company of California to suspend this pipeline by cable along the wall of a cliff. While it was the most expedient thing to do at the time, the arrangement made it difficult for line walkers to check for leaks. *PennWell Publishing Company. Right:* A rotary rig in operation at Goleta. As the bit drilled its way through the earth, the cuttings were carried to the surface by drilling mud mixed from the sacks on the left of the derrick. Once on the surface, the cuttings were screened from the mud, which was returned to the hole. The cuttings were emptied into the slush pit in the lower right. *Security Pacific National Bank Collection, Los Angeles Public Library*

A view of the pioneering offshore well at Summerland, as the abandoned derricks and wharves appeared around 1930. The field, considered the first significant offshore field in the world, was discovered by H. L. Williams in 1886. Piers were built out into the water to serve as drilling platforms. By 1903, 198 wells were producing in this field. *Henry E. Huntington Library and Art Gallery*

Wells of George F. Getty, Inc., in the Summerland Field. George F. Getty was the father and early partner of J. Paul Getty. Note the pipeline in the foreground that is suspended above the high-water mark by posts. *PennWell Publishing Company*

The Rincon Field was another significant coastal field that eventually moved offshore. This is a view of the Richfield Oil Company's Hobson State and Miley-Hobson leases taken from the pier of the Hobson State No. 3. Rotary drill pipe can be seen standing inside one of the derricks to the right of center. *Atlantic Richfield Company*

Workmen building piers and derrick platforms on the coast in the Rincon Field. Note the stack of lumber in the foreground. *Atlantic Richfield Company*

Left: The Rincon Field required oil men to drill not only in the rugged hills near shore, but also on the beach and in the water. Note the stack of drill pipe in the rig in the foreground. These rotary rigs used steam power to rotate the drill pipe with the bit to drill through the earth. As the hole was deepened, a new length of pipe would be screwed into the top of the previous pipe, and drilling would be continued. *Atlantic Richfield Company. Right:* Surefooted workmen constructed a pier at Rincon. *Atlantic Richfield Company*

Another view of the Rincon Field, from the derrick floor of the Richfield Oil Company's Hobson State No. 3 looking toward shore. The elevated walkway in the foreground enabled workmen to reach the well during high tide. *Atlantic Richfield Company*

The derrick base and substructure of Richfield's Hobson State No. 3 were typical of those at Rincon drilled near shore. The rig was elevated above tidal levels. *Atlantic Richfield Company*

The Rincon Field as it appeared in 1938. Most of the wells out in the water still were connected to the shore by piers, but the well in the foreground utilized a free-standing platform. *Standard Oil Company of California*

The Luton-Bell No. 1, discovery well of the Elwood Field, making hole. *Atlantic Richfield Company*

Storage tanks at Elwood. The dikes around the tanks were designed to save as much crude as possible in the event of leaks or fires, which often were combated by shooting holes in the sides of the tanks to reduce the chances that a burning storage tank would explode and spray burning crude over a wide area. *Atlantic Richfield Company*

The Honolulu Oil Company wharf under construction at Elwood. At first the Barnsdall Oil Company refused to allow Honolulu's employees to cross their leases to reach the well site beyond the high-water line. As a result, Honolulu's shifts could be changed only when the tide dropped. Later Barnsdall relented, and regular shifts could be used. *PennWell Publishing Company*

The Elwood Field, sometimes referred to as the Rio Grande Field, like Rincon eventually was moved offshore using the pier-platform technique. One pier reached 2,300 feet into the ocean. Note the huge storage tanks just behind the beach. *Atlantic Richfield Company*

Oil development on the Santa Barbara Mesa in 1934 created this spectacular view. Note the close spacing of the wells. Clouds of steam from numerous boilers powering rigs can be seen. The photo captures in one scene several important elements of life in California in the 1930s—physical beauty, urban development, oil, and agriculture. *Standard Oil Company of California*

Left: Deep-sea diver repairing one of the submarine crude oil pipelines in the Santa Barbara Channel. Once the lines were laid, the only way a leak could be patched was by sending a diver down to work on the line as it rested on the sea bottom. By using acetylene torches to weld, divers could make repairs underwater. *Atlantic Richfield Company. Right:* A Shell Oil Company work-over rig redrilling a well in the Ventura Avenue field in 1934. The well is located on one of the man-made slots, or "islands," on the side of a hill. *PennWell Publishing Company*

Fields in the Santa Barbara–Ventura Basin continued to be productive into the 1940s and 1950s, as secondary recovery techniques were used and as company officials decided to go deeper and discovered new producing horizons. This is a view of Union Oil Company operations at Torrey Canyon in 1953. Note the extensive citrus groves in the background as oil and agriculture continued to coexist. *PennWell Publishing Company*

Operations of Shell Oil Company in Ventura Valley in the late 1940s. Note how "notches" were cut into the hill to the right of center to accommodate drilling rigs. *PennWell Publishing Company*

A Texas Company operation in the South Mountain Field in Ventura County in 1948. The six pumps each represent producing wells, while another is being drilled. Still another well site is being prepared high on the hill in the background. *PennWell Publishing Company*

The Early Boom in the Los Angeles Basin

THE Los Angeles Basin encompasses the region of southern California surrounding the city of Los Angeles. A level lowland plain, the area is approximately twenty-two miles wide and forty-six miles long and lies in southern Los Angeles and northern and western Orange counties. On the north are the Santa Monica Mountains and a line of hills running south and east toward the Santa Ana Mountains. These hills also mark the basin's eastern boundary. Its southwestern border is formed by the Pacific Ocean and the San Joaquin and Palos Verdes hills.

Although the area is relatively small, the Los Angeles Basin contains twenty-eight oil fields, all within thirty miles of Los Angeles. Stretching southward from Venice to San Pedro Bay are the Venice–Del Rey, El Segundo, Lawndale, Redondo-Torrance, and Wilmington pools. The Beverly Hills, Inglewood, Potrero, Rosecrans, Dominguez, Long Beach, Seal Beach, Huntington Beach, and Newport fields run southward in another line between Beverly Hills and Newport Beach. East of Beverly Hills is the Salt Lake Pool, and inside Los Angeles proper is the Los Angeles Field. A final line of discoveries—including the Montebello, West Whittier, Whittier, Santa Fe Springs, Puenté Brea, Olinda, La Habra, East and West Coyote, Yorba, Richfield, and Kraemer pools—stretches southeast of Los Angeles into northern Orange County.

Four of these fields—West Coyote, Montebello, Richfield, and Santa Fe Springs—when they were uncovered in the first two decades of the twentieth century, played a significant role in thrusting California to the forefront of the American petroleum industry. At the time of the West Coyote Field's discovery in 1909, California led all other states in oil production. In 1917, when the Montebello Field was located, it still ranked second. The discovery of the Richfield and Santa Fe Springs pools in 1919 helped the state regain its number-one ranking.

While California's petroleum output was rising during these years, the state's natural gas production also increased dramatically because of the huge production of West Coyote, Montebello, and other Los Angeles Basin pools. Later, between 1928 and 1930, another natural gas boom followed increased production in the Long Beach and Santa Fe Springs fields. Unfortunately, some of the state's greatest wastage of natural gas occurred during these booms. When it became apparent, however, that a valuable natural resource was being squandered, the state legislature enacted the Gas Conservation Act of 1929 to eliminate the problem.

The petroleum legacy of the Los Angeles Basin began in the late nineteenth century. Some of the earliest drilling in the region was inside the City of Los Angeles in 1892. In that year, Edward L. Doheny, who had noticed that local residents were gathering asphaltum for use as a fuel, convinced Charles A. Canfield to sink a mining shaft, using picks, shovels, and a windlass, in the Westlake Park neighborhood on a piece of property they bought for four hundred dollars. The site was near the present intersection of West Second Street and Glendale Boulevard.

At 7 feet they found a seep of light oil. A short time later the oil-soaked shale began to crackle "like popcorn" as natural gas escaped. When they reached a depth of 155 feet, Doheny and Canfield were forced to quit digging because of the fumes. Their attempt to sink a mining shaft temporarily thwarted, Doheny and Canfield utilized a nearby sixty-foot eucalyptus tree, which they sharpened into a makeshift drill bit, to construct a crude cable-tool drilling rig and continued work on the shaft. After forty days they struck oil and natural gas. "I felt like a millionaire," Doheny later recalled.

The two pioneer oil men had found the Los Angeles City Field. Rapidly expanded, by mid-1894 the Los Angeles City Field was producing an estimated thirty-five hundred barrels monthly. Between 1892 and 1900, nearly a thousand wells were drilled in the area. The oil was found at a relatively shallow depth, and most wells were between one thousand and fifteen hundred feet deep.

At first, much of the thick crude was used to pave Los Angeles's streets, but in the summer heat it turned into a sticky goo that clung to anything it touched. Then, in 1902, the Los Angeles City Council agreed to allow crude to be used to heat a local building. When this proved successful, oil quickly replaced coal as the community's chief fuel.

With a ready market, developers continued to expand the pool. Unfortunately, as it reached the community's residential areas, new problems appeared. Houses were surrounded by "chugging and wheezing pumps," gardens and lawns were trampled or flooded, and residents were plagued by lease hounds, a witness recalled. Space was in such short supply that one resident who had a drilling rig in his backyard, when faced with the time it would take to haul away the waste material from the well, simply used his basement as the rig's mud sump.

The noise from the drilling rigs was so great that California's attorney general unsuccessfully attempted to have them declared a public nuisance. "Wells were as thick as holes in a pepperbox," one old-time oil man recalled. In some instances "inexperienced drillers got their wells tangled up with a neighbor's and had to back off and drill again." It seemed as if everyone who could raise the twelve hundred to fifteen hundred dollars needed to drill a well was scrambling for a lease. Within two years of Doheny's discovery, more than eighty wells had been drilled in the area.

Emma Summers, a graduate of the New England Conservatory of Music who had moved to Los Angeles in 1893 to teach piano, had a home near Doheny's first well. Caught up in the oil boom, she was persuaded to invest seven hundred dollars for a half-interest in another nearby well. Drilled by a local hardware man, the well was close to the intersection of Court and Temple streets. When the initial investment was spent, she borrowed additional funds to continue the hole. Eventually she spent several thousand dollars on the well before it was completed as a producer. Afterward she invested in several more holes, always reinvesting the money from each producer in additional wells. Later she bought her own string of tools, hired a drilling crew, and personally oversaw the work on her wells. Although she still taught piano, by 1900 Summers controlled half of the production in the original Los Angeles Field.

At first she sold her crude to local brokers, but as she became more expert in the petroleum business, Summers became a broker herself. She eventually supplied some of the "biggest customers of fuel oil in Los Angeles." Her wells provided fifty thousand barrels of crude per month to several downtown Los Angeles businesses, the Pacific Light and Power Company, and several railroad and trolly systems. Summers became so successful that at the end of the nineteenth century she was known as "California's Petroleum Queen."

Eventually Los Angeles's tremendous production glutted the market, and the price of crude fell to around thirty cents per barrel. Fortunately, at about the same time the several West Coast railroads began to switch from coal, which cost between seven dollars and ten dollars a ton, to the lower-priced oil. By 1903 the Southern Pacific Railroad was purchasing four million dollars' worth of crude annually. This helped stabilize the market, and the boom continued.

Much crude was wasted during the early development in the Los Angeles Basin. One long-time oil man recalled that when he purchased a half-filled oil-storage tank, located near the intersection of Glendale and Beverly avenues, he told the former owner that the tank had to be emptied before he could move it to another site. The former owner replied, "I'll empty it right now," and opened the valves. The oil flowed down the street.

West Second Street "became a greasy, vibrant, oil-soaked little canyon." In some areas, the oil on the ground was so thick that oil-field horses were forced to stand in the crude all the time. Eventually it softened their hoofs so they could not walk. At Echo Park Lake, where it joined what is today Glendale Boulevard, the water and ground became so soaked with oil that in 1907 local oil men set it afire to get rid of the crude. The fire blazed for several days.

The Los Angeles discovery attracted numerous gamblers, prostitutes, and other underworld figures to the strike. Most set up their businesses along Santa Monica Boulevard, near Vermont Avenue. The area quickly became "a raucous oil worker's shantytown." Saloons operated around the clock. "Prostitutes often plied their trade from temporary shelters made from canvas stretched over wooden poles." Usually law enforcement authorities turned their heads on the activity because city officials did not want to disrupt the boom.

Thousands of workers poured into the city. Los Angeles's population jumped from 50,395 in 1890 to 102,479 in 1900, and to 319,198 in 1910. The oil business "is one of the leading industries in the city, and all legislation bearing on it should be liberal," Mayor Meredith P. Snyder declared in 1897.

The original Los Angeles discovery touched off a search for other nearby deposits. The Whittier Field, just east of the city of Whittier, was opened in 1897. The discovery well was the Central Oil Company's No. 1-A, which flowed at a rate of ten barrels daily from 984 feet. By June, 1917, production had reached ninety-six thousand barrels per month; however, by 1938 the field's output had declined to twenty-four thousand barrels monthly, from 163 wells. In 1900, shortly after the original strike at Whittier, the Rideout Heights area of the pool was opened, at the western end of the Puente Hills. Another offshoot of the Whittier find was the Bartolo Area just west of the Puente Hills and about two miles north of the town of Whittier.

In 1919, near La Puente, on the Didier Ranch, J. Paul Getty drilled his first well in California. Unfortunately, it was "a dismal failure." As Getty recalled, within seven months he had spent "nearly $100,000, and the drillers had managed to get down only 2,000 feet." Although his first individual effort failed, Getty continued to pioneer California oil-field development, both as a partner with his father, George F. Getty, and on his own, until he had built the Getty Oil Company into one of the world's major energy concerns. J. Paul Getty's involvement in the oil industry had started in 1903, when his father formed the Minnehoma Oil Company. The younger Getty purchased one hundred shares of Minnehoma stock early in 1904, at a cost of five dollars a share. In the following years, Minnehoma was active in several Oklahoma oil fields. During that time, the younger Getty was trained as a driller and purchased individual leases, for which his father furnished the money for a 70-percent share.

In 1906, the Gettys moved to California, first to San Diego and later, in 1909, to Los Angeles. In 1916 J. Paul Getty entered the petroleum industry on a fulltime basis, was named one of Minnehoma's directors, and incorporated the Getty Oil Company, in partnership with his father. The elder Getty was named president and treasurer of the new firm, with 70-percent ownership, and his son was named secretary, with 30 percent of the stock. In 1923, J. Paul Getty inherited both the Minnehoma Oil and Gas Company and George F. Getty, Inc. Because his father's companies were, like his own, "primarily engaged in oil exploration and production," he combined their holdings with those of Getty Oil Company.

In the western part of the Los Angeles residential district, the Salt Lake Oil Company uncovered the Salt Lake Oil Field in 1903. Between the time of its discovery and 1915, the pool was hurriedly developed by three major concerns: the Salt Lake Oil Company, the Arcturus Oil Company, and the Rancho La Brea Oil Company—all operated by the Associated Oil Company. Eventually 350 wells were drilled in the field's one thousand productive acres, which stretched north, east, and west of the La Brea fossil pits north of Wilshire Boulevard and between Highland Avenue on the east and San Vicente Boulevard on the west. Peak production was reached in 1908, with 185 wells flowing at 4,535,800 barrels of crude.

Located just to the west of the Salt Lake find, the Beverly Hills Field was uncovered in 1908. It reached its peak in 1912, with an annual production of 246,223 barrels of crude. However, it proved to be a relatively small find, with an average daily output of thirty-five barrels per well.

The extensive drilling in the metropolitan Los Angeles area eventually created many problems, as local citizens became more aware of noise pollution and the unsightliness of neighborhoods. Early efforts to curb the drilling excess met with failure. When the Los Angeles City Council restricted the drilling of oil wells in residential areas, oil men simply applied for permits for drilling water wells and then claimed that while drilling the water wells they had accidentally struck oil. In the 1940s oil men began to hide their drilling rigs in isolated canyons away from residential areas, but as the community grew, even these once-remote regions were covered with houses.

The matter came to a head with the opening of the Sansinena Field near Whittier in May, 1945. Local citizens had insisted on strict zoning regulations to limit drilling sites, and many residents protested any variance in the rules. To solve the problem, Sansinena's operators agreed to soundproof their rigs by covering them with two layers of a "vinyl-coated glass cloth with one-inch sheet fiberglass filling" that was specially "heat-processed and quilted." In addition special precautions were taken to ensure there would be no leakage of fluids.

It proved to be a workable arrangement that satisfied local residents and, at the same time, because of the bright yellow color of the inside of the soundproofing, allowed drilling crews to work twenty-four hours a day in "daylight." Later the rigs were even camouflaged to blend in with their surroundings. Eventually, that technique was extended to offshore operations as well.

Following Salt Lake, the next major discovery in the Los Angeles Basin was the Greater Coyote Hills Field, uncovered on April 26, 1909, in the extreme northern portion of Orange County, about six miles southeast of the Santa Fe Springs Field and three miles northwest of Fullerton. It was the first discovery in the Los Angeles basin that was not made because of a nearby surface seepage of oil or gas. Instead, the Murphy Oil Company started the discovery well, the Coyote No. 3, after learning that a local water well had produced a show of oil. By June, 1918, the output of the West Coyote Hills Pool was 31,600 barrels daily, and eventually the field was expanded to include a thousand acres.

West Coyote was different from many other early California pools. Because all of the field, except a small portion on the edges, was leased by Standard Oil of California, there was no mad scramble to develop the area. Instead, the lack of competition allowed

Standard Oil to drill only when it was necessary, and the pool was developed by the most economical means possible. The discovery of oil where there were no surface seeps made many California oil men bolder in their wildcatting projects. Later the East Coyote, Richfield, Dominguez, and Long Beach fields were uncovered, and none of them were near oil seeps.

About two miles northeast of Fullerton and three miles east-southeast of the West Coyote Field, the East Coyote Pool, originally known as the La Habra Field, was located in 1911. Anaheim Union Water No. 1, the discovery well (belonging to the Amalgamated Oil Company, which later became a part of Tide Water Associated Oil Company) caused quite a stir as one of the first rotary wells in the region. It flowed at six hundred barrels daily from between 2,830 and 3,340 feet.

The Yorba Linda Pool of the Greater Coyote Hills Field was located in 1937 by the L. S. Simmel oil company, which later became the S. V. Smith oil company. The discovery well, the Todd No. 1, originally flowed at 150 barrels per day, but within thirty days of its completion, it had declined to 20 barrels per day. As a result oil men took little interest in the region until the Shell Oil Company completed its Olinda Land Company F Well in December, 1937, a 1,200-barrel-per-day producer. Afterward twenty wells were drilled in a 120-acre area. Yorba Linda's output was never great, and production peaked in July, 1938, at 700 barrels daily.

The Richfield Oil Field in northern Orange County was composed of two distinct pools: Kraemer and Richfield. The Kraemer Pool, also called the Santa Ana Canyon Field, was located on September 16, 1918, by the Standard Oil Company. Flowing at 176 barrels of water and 144 barrels of oil daily, the discovery well, the Kraemer No. 1, was completed at a depth of 2,762 feet. Eventually twenty-five wells were drilled in the pool, with seventeen completed prior to 1923. As of 1923 the field had expanded to approximately 120 acres. Individual well production varied from a minimum of 5,000 barrels to a maximum of 440,000 barrels by June, 1939. As of June 1 of that year 1.4 million barrels of oil had been produced in the Kraemer Pool.

The Richfield Pool was uncovered on March 11, 1919, by the Union Oil Company, about six and a half miles east of Fullerton in Orange County. Showings of natural gas in the low-lying hills had prompted the drilling of a wildcat well, the Chapman No. 1, by Union. The wildcat came in from 3,025 feet as a 1,750-barrel-per-day producer. A deeper zone was discovered by the Standard Oil Company on June 22, 1920, when its Kraemer No. 2-6 flowed at 185 barrels daily from 4,130 feet—a rate of flow that was maintained for the next twenty-two years.

Although there was a lower demand for oil in the post–World War I era, the Richfield Pool was rapidly developed. Approximately three hundred wells were drilled on thirteen hundred acres in the region. Many of them maintained a constant flow for years, and a cumulative output of between nine hundred thousand and two million barrels per well was not unusual. As of March, 1942, total production from the pool was approximately ninety-five million barrels of crude. In addition the oil was extremely rich in casing-head gasoline, averaging 1.4 gallons per thousand cubic feet of natural gas, and the ratio of natural gas to oil stood at approximately one thousand cubic feet per barrel.

To the east of Los Angeles and slightly west of the San Gabriel Fault, along the western edge of the La Merced Hills, the Montebello Field was uncovered by the Standard Oil Company on February 28, 1917. The discovery well, the Baldwin No. 1, was originally spudded in with a rotary rig on December 6, 1916, but was completed with cable tools. Eventually, the pool's 980 productive acres stretched from the southwest end of the La Merced Hills eastward into the valley of the Rio Hondo. Total cumulative production as of October 1, 1940, stood at approximately 104,750,000 barrels from 214 wells.

The West Montebello Pool was uncovered along the southwestern edge of the field when the Kern Oil Company completed its Monterey No. 20 on September 28, 1936. Eventually eight productive horizons were uncovered in the field, which stretched for a mile and a half inside the city limits of Montebello. On January 1, 1940, 160 wells were producing 83,700 barrels of crude daily.

Oil men had long known about the presence of natural gas in water wells southeast of Los Angeles on the east side of the San Gabriel Fault. As early as 1907 the Union Oil Company drilled in the region, which later became the Santa Fe Springs Field. Union's early wells were abandoned because of mechanical difficulties, however, and the company did not resume its attempt to develop the region for another decade. In February, 1917, the Union Oil Company spudded another well at Santa Fe Springs and completed it two years later. Although the second Union well's initial production was three thousand barrels daily, its output quickly dropped to between one hundred and one hundred fifty barrels a day. Oil men paid little attention to the discovery until November, 1921, when another Union Oil Company well, the Bell No. 1, came in at 2,588 barrels per day. Afterward Santa Fe Springs was rapidly developed. In 1923 and 1929 the field produced more oil than any other California pool.

Among the earliest developers of Santa Fe Springs were George F. and J. Paul Getty. Two days after the Bell No. 1 was completed, they drove through the area examining the topography. They were accompanied, as the younger Getty later recalled, by a "highly regarded and high-priced geologist." Although the geologist believed that "the most promising oil land lies to the east of the [Bell] well," J. Paul Getty, who was driving, turned southward on Telegraph Avenue. "The land around here doesn't look as though it would produce very much oil," the geologist commented; "I'm sure the pool lies to the east."

"Shall we take another look in that direction?" Getty asked his father. "Not yet," was the reply. "As long as we're here, we might as well drive a little farther. I'm not so sure that the best oil land isn't here in this section."

As J. Paul Getty recalled: "A few moments later, we saw a long, heavily loaded freight train straining and laboring across what appeared to be a level expanse until it reached the grade-crossing at Telegraph Avenue. After that, the locomotive's power obviously eased off, but the train nonetheless began to gather momentum and pick up speed. This could mean only one thing: the terrain was not as flat as it seemed to be." "Gradients," he explained, "imperceptible to the naked eye, sloped away from a Telegraph Avenue summit."

"Did you notice that train, Paul?" the elder Getty asked. "I'll say I did!" was the answer. "The top of the structure—the dome—is right here, along Telegraph Avenue," George Getty continued. "The richest part of the pool extends here, to the south of Union Oil's well." Shortly afterward, the elder Getty purchased the Nordstrom Lease, four lots 50 feet by 145 feet, fronting Telegraph Avenue. The total price was $693. On August 26, 1922, Nordstrom No. 1 was completed as a 2,300-barrel-per-day producer. Within the next seventeen years, the lease produced crude worth over $6 million.

Santa Fe Springs was a flush production field, with a tremendously high initial output of crude. Once the rush started, thousands of oil men poured into the vicinity. As in all boom towns, space was at a premium in Santa Fe Springs. There were more workers and men than available accommodations. The Getty Oil Company's office "replaced a Japanese cabbage patch among the surrounding market gardens." One of the best-known regions at Santa Fe Springs was "Hell's Half Acre." In it wells were "drilled so close together that the derrick legs interlocked and men walked from one rig's platform to another without touching the ground." It was not uncommon on windless days for the steam from the thousands of boilers in the field to cover the pool with "man-made fog" so thick that it was difficult to see what was happening on nearby wells.

Alphonso E. Bell held title to approximately one hundred fifty acres in the center of the pool, more than anyone else in the field. At the height of production, Bell was receiving a reported one hundred thousand dollars a month in royalty. His total earnings for his property were estimated at between six million and twelve million dollars, depending on the price of crude.

Santa Fe Springs' monthly production peaked in August, 1923, at 322,500 barrels of crude daily. Production for that entire year also was a record—80,671,112 barrels. So great was the flow of oil from the field that the price of crude throughout the country began to be affected. The Pacific Coast market simply could not absorb the tremendous output. During the summer of 1923, intercoastal shipments of oil, most of which originated at Santa Fe Springs, reached two tankers per day, or a total of 15,241,621 barrels for July through September.

Production jumped again in July, 1928, when the Wilshire Oil Company completed its Buckbee No. 1 and uncovered a deep producing horizon at 5,860 feet. Flowing two thousand barrels per day, the Buckbee No. 1 touched off another drilling rush in the pool. The following year, activity at Santa Fe Springs resembled the earlier boom. In conjunction with the deep pay discovery, Santa Fe Springs' production of natural gas was greatly expanded between 1928 and 1930. By late 1928 the pool contained 23 wells flowing at seventy-five thousand barrels daily from the deeper production, and another 215 wells were pushing holes toward the deep production.

The tremendous natural gas pressure at Santa Fe Springs posed many problems. Of special concern were the blowouts that occurred whenever a well encountered an unexpected pocket of high-pressure gas. The possibility of a sudden blowout was always on the minds of drilling crews in the area, and it was an understandable concern. In January, 1922, for example, the Union Oil Company's Alexander No. 1 unexpectedly struck such a pocket at 2,060 feet. Mud began to boil out of the casing, and the ground began to roar. As the gas blew the drill pipe out of the hole, the drilling crew fled for their lives to escape the falling debris. One roughneck was on the derrick when the gas blew in and, as a column of rocks and mud was thrown over the crown block, he jumped from the rig, dropping thirty-five feet into the slush pit. Although covered with mud and crude, he scrambled to safety.

One of Santa Fe Springs' most spectacular fires was at the Getty Oil Company's Nordstrom No. 17, which burned for six weeks in early 1929. The well had blown out before drilling had been completed, and the escaping natural gas quickly ignited. "When the boiler plant caught fire," Frank Whitty recalled, "a solid stream of gas and oil poured over the derrick, which melted." To fight the fire, a crew under Whitty "went back thirty yards to try to tunnel to the casing to draw off the pressure." Once the pressure was eased, the surrounding area was cleared of surface oil to keep the flames from spreading. Even so the Nordstrom No. 17 "blew for nearly two months" before being brought under control.

Santa Fe's output reached 77,576,147 barrels of oil in 1929. Afterward production again declined. Many of the older wells were plugged back to more shallow producing horizons. By January, 1940, Santa Fe Springs' production stood at 36,064 barrels daily from 658 wells. Cumulative production as of December 1, 1939, was 458,397,637 barrels, making the field at that time the third greatest petroleum producer in California, ranking behind only the Los Angeles and Midway-Sunset discoveries.

The extensive development of the Los Angeles Basin prior to the 1920s helped supply the tremendous demand for oil brought on by World War I. During the war many of the wells were pumped around the clock, and some of the older discoveries were simply exhausted. Because of the overproduction, in 1919 California was struck by a severe gasoline shortage, alleviated only by rushing "millions of gallons" of gasoline to southern California from Wyoming and Texas.

Prior to the Roaring Twenties, the Los Angeles Basin had produced four of the nation's greatest oil fields. The fields and their ranking among the nation's producing pools at that time were: Santa Fe Springs, fifth; West Coyote, thirty-ninth; Montebello, forty-fifth, and Richfield, sixty-eighth. Cumulative production from these fields at the end of 1949 was 958,401 barrels of crude. As late as 1980 the original Los Angeles field still had almost forty producing wells, flowing at about 50,000 barrels of crude annually.

A Columbia drilling rig in operation at the intersection of Glendale Boulevard and Colton Street in Fullerton, in 1890. This was one of the early successful operations in the Los Angeles Basin and a masterpiece of simplicity. Power was supplied to the walking beam by a cable stretching from a central power station to the back of the post supporting the beam. As the cable moved back and forth, so did the post. This motion in turn was transferred to the walking beam, which raised and lowered the bit in the hole. Notice that the men on the derrick floor are posing before cameras, undoubtedly for a publicity photograph to be sent to prospective investers. *Henry E. Huntington Library and Art Gallery*

A panorama of the Los Angeles City Field in 1895. By this time the "wells were as thick as the holes in a pepper box," with many derricks crowded against the houses. *American Petroleum Institute*

One of Edward L. Doheny's many wells in the Los Angeles Basin area. Doheny may be seen standing in the center of the photograph between the boiler and the derrick. *Henry E. Huntington Library and Art Gallery*

The intersection of Court and Toluca streets in Los Angeles around the beginning of the twentieth century. By this time nearly a thousand wells had already been sunk in the pool. The structure in the center of the photograph houses a large boiler and engine used to power nearby wells. Notice the multitude of cables radiating from the structure. Power generated by the engine was transferred by the cables to the wells. *Henry E. Huntington Library and Art Gallery*

By 1901 the wells in the Los Angeles region had penetrated the residential areas of the city. Many of these wells are literally in the backyard of the nearby homes. The absence of boilers and cable tools on the rigs indicates that these wells have already been completed and, instead of removing the derricks, the oil men have simply left them standing. *Henry E. Huntington Library and Art Gallery*

Canfield & Clampitt Oil Company wells in Los Angeles in 1901. Most of these wells have already been completed and the derricks left standing. A network of above-ground pipeline connects the wells with the storage tanks in the right foreground. Oil was so plentiful that it was sometimes easier to allow excess oil to flow down the street than to empty the tanks prior to their removal. *Henry E. Huntington Library and Art Gallery*

A view from the center of the intersection of Wilshire and Carson streets in Los Angeles. Obviously a well-developed residential area with paved roads and streetlights, the region was nevertheless invaded by oil men. *Security Pacific National Bank Collection, Los Angeles Public Library*

The Pina No. 1 in the Los Angeles Field. Notice that a wall of bricks has been built around the well's boilers. Undoubtedly the idea was for the bricks to provide protection should the boilers explode, but it is more likely that if the boilers had exploded the bricks would have been thrown like shrapnel over a wide area and would have presented more of a danger then a safety factor. *Standard Oil of California*

A well drilled in the middle of a Los Angeles street during the early oil boom. Apparently the well was drilled before the street was constructed because it stands on land that has not been leveled. A blinking light has been placed in the fence in front of the well to warn oncoming motorists. Barely visible at right, a Richfield Oil Company service station offers gasoline for eight cents a gallon. *Security Pacific National Bank Collection, Los Angeles Public Library*

Standard Oil of California's Emery Camp in the Whittier Field in 1912. Every effort was made to make the camps attractive to employees. Notice the all-around porch surrounding each building. Behind the second house from the right a well-maintained outdoor privy can be seen. *Standard Oil of California*

As early as 1909 the greater Los Angeles area was becoming a major refining center. In that year T. A. Winter, J. R. Jacobs, and George Gillons organized the Los Angeles Oil & Refining Company, which had a daily output of eighty barrels. It was later acquired by the Richfield Oil Company. The large building on the left was the original headquarters of the Los Angeles Oil & Refining Company facility, as it appeared in 1926. *Atlantic Richfield Company*

Boust City near Los Angeles was one of many oil towns that were established during the Los Angeles Field's boom years. At the time of this photograph streets had already been laid out, but Boust City's biggest attractions were still one hotel and the Eagle Bar and Grill. *American Petroleum Institute*

Left: George F. Getty, the father of J. Paul Getty, moved to Los Angeles with his son in 1909. Eventually the two formed a partnership and organized the Getty Oil Company. George F. Getty held 70 percent of the stock and was named president. *PennWell Publishing Company. Right:* His son, who held the other 30 percent of the stock, was named secretary. The younger Getty drilled his first California well near Puente in 1919. *Getty Oil Company*

The General Petroleum Company's lease near Brea in the Los Angeles Basin area. The wells have been drilled in lines along the ridges running down from the crest of the main ridge toward the narrow stream in the foreground. It was difficult to build derricks on the uneven ground of the ridges, but, because of the danger of flash floods down the arroyos, most oil men preferred to construct their rigs on the high ground. *Henry E. Huntington Library and Art Gallery*

Interior view of a rotary rig operating on the Rancho La Brea. The two men in the center are attaching another string of pipe to the bit in the hole. *Security Pacific National Bank Collection, Los Angeles Public Library*

A huge tank farm complex was developed by Union Oil Company of California at Brea to handle the tremendous outpouring of crude from nearby wells. Most of the larger tanks were concrete-lined excavations that were covered with steel tops. Outside the earthen embankment around each concrete tank was a second terrace that served as a fire wall to contain any flaming oil should the tank catch fire. *PennWell Publishing Company*

Union Oil Company's Brea tank farm ablaze in the spring of 1926. While the double terracing did contain the flames somewhat, when this picture was taken the fire had already spread to two tanks and adjoining grassland. *PennWell Publishing Company*

An aerial view of the Brea Field in the center of the Los Angeles Basin. Notice that the pool is surrounded on all sides by residential areas that extend to the very limits of the field. The pool appears as a plot of undeveloped property amid a thriving city. In the upper background is Hancock Park. *Atlantic Richfield Company*

This aerial view of the Montebello Field shows a mixture of newer steel derricks with older wooden structures. Most of the derricks have been built on high ground to avoid sudden floods. Note the maze of roads connecting the numerous rigs and how the derricks have moved off the ridge and into the orange groves in the right of the photograph. *Standard Oil of California*

By the end of the 1920s, California had established itself as a major supplier of crude oil and a center of America's petroleum industry. Because of this, it hosted the Oil Equipment and Engineering Exposition in 1930, 1931, and 1932. However, during the Great Depression the show was moved to Houston, Texas, and renamed the Oil World Exposition. *Security Pacific National Bank Collection, Los Angeles Public Library*

Right: A soundproof rig operating in the Beverly Hills Field. In the background is a 20th Century–Fox movie set. The initial effort to lessen noise pollution came in the Beverly Hills Field in the 1940s. The necessity of eliminating outside noise while shooting movies added to the cry for soundproofing by area residents and businesses. The pumping unit in the foreground is pulling oil from one of the wells completed in the first decade of the twentieth century. *PennWell Publishing Company. Left:* When the Sansinena Field was opened in May, 1945, local zoning regulations required that all rigs be soundproofed. To accomplish this, oil men covered their rigs with two layers of vinyl-coated glass cloth with fiberglass filling. *PennWell Publishing Company*

A rotary rig operating in the Montebello Field in the mid-1930s. Note the pipe stacked along the derrick. It has been removed from the hole by the drilling crew in order to change the bit. Once the bit has been changed, the pipe lengths will be reconnected and lowered back into the hole. *American Petroleum Institute*

There were no roadways to many of the early Los Angeles–area camps. At this Richfield Oil Company camp at Signal Hill, tire tracks seem to converge from all directions. *Atlantic Richfield Company*

The Getty Oil Company No. 17 afire at Santa Fe Springs on September 16, 1928. Both J. Paul Getty and his father George Getty played prominent roles in the development of the field. A sign on the store to the right of the photograph offers malted-milk shakes to those working on the drilling rigs. On the left a fire hose is spraying water on nearby structures to keep them from catching fire from flying sparks. *Cities Service Oil Company*

A cratering well at Santa Fe Springs. Because of the field's huge production of natural gas, fires and craters were common occurrences. This crater, formed by natural gas blowing out around the well-head, did not ignite; however, little remains of the drilling rig. *Standard Oil of California*

The remains of the explosion and fire of the Rhodes well at Santa Fe Springs. Although the storage tank near the well, unlike the sheet-iron boiler house in the right foreground, was strong enough to escape the explosion, it was melted by the heat of the flames. Note the mixture of wooden and steel derricks in the field. *Standard Oil of California*

A view of one of Santa Fe Springs' many fires, from the crown block of a nearby well. The fire, from a runaway gas well, has already destroyed the wooden drilling rig and, because there is no wind, is shooting flames straight into the air instead of toward neighboring wells. *Atlantic Richfield Company*

The wells were so close together at Santa Fe Springs that any fire threatened to destroy the entire field. In this explosion in June, 1929, the force of the blast blew down most nearby structures, making it easier for firefighters to control the flames. Notice that despite the fire the boilers on other wells nearby are still operating as the wells continue to make hole. *Atlantic Richfield Company*

This fire resulting from a blowout at Santa Fe Springs has melted the well's steel derrick, which is crumpling over the well hole. The water in the foreground has accumulated as firefighters sprayed nearby wells and buildings in an attempt to prevent the flames from spreading. *Standard Oil of California*

To allow greater access to Los Angeles' crude the harbor facilities at nearby San Pedro were developed to handle a large volume of tanker traffic. Much of the work was undertaken by Union Oil Company of California, which financed this project in 1890. Large boulders would be piled offshore to form breakwaters or used to form the foundations for piers in the harbor area. *Union Oil Company of California*

A Christmas tree rig on a Richfield Oil Company well at Santa Fe Springs. Designed to prevent blowouts, such equipment lessened the dangers of drilling in the field. *Atlantic Richfield Company*

Because of the great production of the Los Angeles Basin, the region became a major refining center. Located in the center of surrounding orange groves, this is the Standard Oil Company of California's El Segundo refinery complex in 1911. Notice that many of the surrounding wells have coverings between the drill floor and the first horizontal beams of the derricks. These were designed to prevent any oil from spilling into the orange groves. *Standard Oil Company of California*

Union Oil Company's first service station was opened at the intersection of Sixth and Mateo streets in Los Angeles in 1913. It was the forerunner of thousands of such facilities that appeared almost overnight as the automobile age began in America. For seventy-five cents the attendant on duty would also waterproof an automobile top. *Union Oil Company of California*

One of the area's largest refinery complexes was the Richfield Oil Company Refinery located at Watson just inside Los Angeles County. This is an aerial view of the first crude battery completed at the refinery complex in 1923. Notice the huge storage tank in the upper right. Although it appears to be an earthen tank, it is actually concrete-lined. Four smaller, 55,000-barrel steel storage tanks, some of which are still under construction, are on the left. *Atlantic Richfield Company*

Teams of mules pulling scrapers carry dirt to waiting wagons during work on the storage tank at Watson. The lights hung from the poles allowed work to continue around the clock if necessary. When completed, the tank was one of the largest of its kind on the West Coast. *Atlantic Richfield Company*

At the time of its completion, Richfield's Watson refinery was one of the most modern in the world and incorporated the latest assembly-line techniques. This worker is operating a mechanical case-top nailing machine, which automatically attached the tops to cases of oil products as they neared the end of the assembly line. Afterward the cases were carried to a loading dock by the line of rollers on the right. *Atlantic Richfield Company*

The Richfield Oil Company made every effort to keep its Watson refinery in the forefront of technological advances in the petroleum industry. These 50,000-barrel combination topping and cracking units produced "Miracle Hi-Octane" gasoline as early as 1939. *Atlantic Richfield Company*

The Roaring Twenties in the Los Angeles Basin

At the beginning of the Roaring Twenties, California was producing 103,377,000 barrels of crude and 58,567,772,000 cubic feet of natural gas annually. The total value of the state's petroleum had reached almost two hundred million dollars per year by the start of the decade, and 1920 witnessed a resurgence of drilling in the Los Angeles Basin that made California the "king of the oil states." During this decade, California ranked first among the petroleum producing states in 1923, 1924, 1925, and 1926; it held second place in 1920, 1921, 1922, 1927, and 1929 and third place in 1928.

Also during the Roaring Twenties, six of America's greatest oil fields of the first half of the century were uncovered in the Los Angeles Basin: Huntington Beach in 1920, Long Beach in 1921, Torrance in 1922, Dominguez in 1923, Inglewood in 1924, and Seal Beach in 1926. They ranked eighth, third, fifty-seventh, thirty-fourth, thirty-sixth, and sixty-third, respectively. By the close of 1949 the total cumulative production of these fields would reach 1,780,593,000 barrels of oil.

The boom years of the 1920s started at Newport Beach, just inland from Newport Bay. Although the local tar sands and oil seeps were well-known, it was not until 1903 and 1904 that oil men began to examine the area. The first attempt was made by the Walker Brothers Oil Company, and, although their effort was a dry hole, for the next fifteen years wildcatters continued to drill along the Newport Mesa and tidal flats on the west end of Newport Bay. Finally, in 1922 and 1923 the Fulkerson group drilled a tidal flat well, the Fulkerson No. 1, that flowed at fifteen barrels daily from 775 feet. This was the discovery well of what became the Newport Field.

In early 1925 Barnett Rosenberg completed the Mesa No. 1 in the mesa area to the north of the tidelands, producing 185 barrels per day. This opened an entirely new area to oil men, and a short time later the Julian Oil Company completed a well nearby with an initial flow of 1,200 barrels daily. After eight months, though, its production fell to 32 barrels of crude and 40 barrels of water daily. Additional deep drilling proved unsuccessful.

Throughout this period the Newport Field proved to be a disappointment to oil men. Although most wells had an initial strong showing of oil, the rapid incursion of water and the poor quality of crude produced little profit. Little oil came from the pool after 1931, when most of the wells were abandoned. By November of 1940, Newport's cumulative output stood at 154,000 barrels of crude.

Although Newport did not touch off the second Los Angeles Basin boom, it did keep oil men interested in the region, and shortly afterward the tremendously rich Huntington Beach Field was uncovered. One of the most complex fields in the Los Angeles Basin, the Huntington Beach Field is located about forty-one miles southeast of Los Angeles, just north of the town of Huntington Beach. Oil men first began to notice the area when local water wells encountered natural gas (eventually the field became a major source of natural gas in California). Standard Oil Company of California completed the pool's discovery well on May 24, 1920, at a depth of 2,199 feet. Although the initial producer flowed at only forty-five barrels per day, shortly afterward, on November 13, the Standard Oil Company brought in its Bolsa Chica No. 1, with an initial flow of two thousand barrels daily. This touched off a rush to the area.

Eventually, the Huntington Beach Field was developed as four separate pools: the Old Field, the Main Street Field, the Surf Area, and the Townsite Tideland Area. Deep production in the Old Field, which stretched along the area's main fault line, began on April 24, 1921, when the Eddystone Oil Company, later acquired by Shell Oil Company, completed its Ashton No. 1, which from a depth of 3,445 feet flowed at thirteen hundred barrels daily. The pool was California's first town-lot discovery, and most of the region was developed by smaller oil companies. As a result, the well spacing was close.

Originally known as the Barley Field prior to 1924, the Townsite Tidelands Area, south of the Old Field, was first explored, unprofitably, by the Standard Oil Company. When the Wilshire Oil Company completed its H. B. No. 1 on September 14, 1926, with an initial flow of 700 barrels per day, extensive development of town lots started. Most oil men quickly realized that the pool extended offshore, and with the advent of directional drilling, or whipstocking, they began to examine previously unreachable parts of the field.

The first whipstocked well was the McVicar and Rood Oil Company's Vicaroo No. 1; whipstocking allowed the hole to be started onshore and then slanted offshore by the use of a beveled bar of steel placed in the bottom of the hole. Vicaroo No. 1 was completed on January 1, 1933, with a strong flow of natural gas and forty-eight hundred barrels of oil. Some ninety wells were whipstocked offshore before the practice was halted by legal action.

One section of the Huntington Beach Field, which became known as the Encyclopedia Section, was a nightmare for lease hounds. The region acquired its name in the early twentieth century when a small acreage of unpopulated land overlooking the Pacific Ocean was acquired by a land promoter who subdivided it into twenty-foot by twenty-five-foot lots for sale to potential homeowners. When the California real estate market entered a slump shortly thereafter, the lots became impossible to sell. The enterprising land promoter then joined with a New England–based encyclopedia company to offer the land as a bonus for individuals who purchased sets of encyclopedias. A massive sales campaign resulted in the disposal of most of the lots in this fashion. However, the majority of the buyers were more interested in the books than in the land, and many of the property deeds were filed away and forgotten.

After the opening of the Huntington Beach Field, it was discovered that the Encyclopedia Section contained some of the most valuable real estate in the area. A frantic search began to find the property owners, many of whom lived on the East Coast. Many had neglected to pay taxes on the property, and the land had been resold for as little as one hundred dollars a lot. However, the original purchasers who had maintained their titles suddenly discovered that they owned rich oil lands worth hundreds of thousands of dollars.

Close on the heels of the Huntington Beach strike came the discovery of the Long Beach, or Signal Hill, Field. Located about twenty miles south of Los Angeles, just north of San Pedro Bay and inside the city limits of Long Beach, Signal Hill received its name because the Spaniards had signaled ships at sea from its summit since it was the highest point in the area. In the late 1800s the town of Signal Hill was started on the rise. In the post–World War I era of land speculation, many residential lots had been sold on the "scenic front side of the hill."

Oil men first began exploring the region in 1916, when the Union Oil Company completed its Bixby No. 1 just to the southeast of the intersection of Wardlow Road and American Avenue. At that time most of the area was still farm land. Plans had been made for residential development, and a few houses had already been built, but little more had been done than laying out streets and subdivisions.

When the Bixby No. 1 did not prove commercially successful, many oil men concluded that the region was not productive. One disgruntled Union Oil executive declared, "I'll drink all the oil there is at Long Beach." Yet some remained convinced that there was oil nearby. In 1918, Frank Hayes, a Shell geologist, recommended to his superiors that the firm lease the area but was overruled by Wilhelm van Holst Pellekaan, head of the company's geology department. In April, 1920, Alvin Theodore Schwennesen assumed command of Shell's geologists and, after reviewing Hayes's original proposal, convinced the company to lease 240 acres at Signal Hill for sixty thousand dollars.

The discovery well, the Alamitos No. 1, was spudded in just below the projecting upper part of a hill at the northeast corner of Hill and Temple streets on the edge of a lease acquired from the Alamitos Land Company. When Schwennesen's boss discovered the plan, he reportedly declared, "That damned fool Schwennesen has got us drilling a well in Long Beach, and I've got to get out there and stop it." But it was too late to halt the venture; work was started with a rotary rig on March 23, 1921. On May 2 the hole reached 2,765 feet and gave a showing of oil. A cable tool rig took over, so that the mud from the rotary rig would not seal off the oil sands. Shortly afterward nearly seventy feet of oil was found in the bottom of the hole. Later, as the driller, O. P. "Happy" Yowells, was bailing out the hole, the crew heard "a growl below like waves roaring through a sea cave," but much to everyone's disappointment no oil gushed from the wellhead.

The disappointment did not last long. On June 23, at 9:30 P.M., the Alamitos No. 1 blew in, with so great a gas pressure that the crude rose to a height of 114 feet before the hole sanded up. After two days of cleaning, the well started producing again, with an initial flow of 590 barrels daily. Production quickly rose to 1,200 barrels a day. Eventually the Alamitos No. 1 produced 700,000 barrels of oil.

With the oil came a huge flow of natural gas. Shell hurriedly started several other wells in the area. Its second completion was a gasser that blew out and caught fire. Since it was flowing at an estimated twenty million cubic feet per day, it took the combined power of seventeen steam boilers working three days to extinguish the flames. Another Shell well, the Martin No. 1, also blew in as a gasser and caught fire. It took one hundred pounds of dynamite to blow out that fire.

One of the most spectacular fires at Signal Hill was the Shell Oil Company's Nesa No. 1. Ironically, every precaution had been taken to prevent a blowout. After the hole had been drilled as deep as possible with a rotary rig, it was cemented to hold down any unexpected natural gas flow. Two weeks after the cement had been poured, on September 1, 1921, the cable tool crew drilled through the hardened cement at the bottom of the casing and completed the well. During the delay a pocket of natural gas had collected below the casing, and when the cement plug was penetrated the released gas shot 125 feet into the air. Almost immediately the escaping gas caught fire. It was two days before enough equipment could be gathered to extinguish the flames.

"Blowouts, gassers, gushers, fires, explosions, craters, geysers" were all part of Signal Hill's development. Such instances caused a great amount of damage, one magazine reported. "Machinery, trees, telegraph poles and buildings are torn to bits and scattered with a sea of mud over large portions of the landscape." One well that blew out at Signal Hill covered a house two hundred feet away with "mud, oil and rocks, ruined the trees in the yard, bespattered and half wrecked the family automobile, and left the lawn covered with an eight-inch coat of oil, mud and rocks."

Much of the region's natural gas was flared off, and the burning gas often made it so light at night that the drilling crews needed no other illumination to work by. Not all was burned, however. Much raw gas was spewed into the air, to become a time bomb when the atmospheric conditions were just right. During the winter, the cold fog held the natural gas fumes to the ground, where the slightest spark could result in an explosion. Automobiles often touched off the gas that gathered under viaducts, and the open flames on the many steam boilers and heating stoves would ignite fumes.

Some of the gas was piped to market, and the field was inlaid with a network of pipelines. Many people living along the pipeline right-of-way looked upon the pipelines as a ready source of natural gas and helped themselves. Indeed, illegal taps were commonplace; the Signal Hill Gasoline Company once discovered that a cafe and two boarding houses were drawing natural gas from its lines. While many tapped the gas pipelines for light and heat, others wanted to draw off "drip" gasoline, which could be burned by automobiles. A condensate, the drip was collected in holders placed at various intervals along the natural gas pipelines. One enterprising individual tapped a gas line that ran under his home, built a still in his kitchen to make gasoline, and supplied several service stations with his product.

To many visitors the large number of burning flares "fostered a carnival atmosphere" in the pool. Many oil promoters took pains to cultivate such a feeling. Arrangements were made to provide free bus rides to Signal Hill for those wishing to visit the latest oil strike. So many people crowded around the Alamitos No. 1 that the crew was forced to run them off the drilling platform with axe handles. Eventually a tall barbed-wire fence was constructed to keep away the sightseers.

"Large red and white circus tents," some of which had previously been used for religious revival meetings, were erected to house prospective investors. "Usually the promoters would hire a barker to stand outside and pull in customers." Once inside, the people generally were fed a free meal, and then the promoter made his sales pitch. It was not uncommon for "smooth talking promoters" to auction off oil leases "from the back seats of their open cars."

Apparently it worked, for as one witness recalled, "people was flocking to the counter table to shell out their $100 bills." Some of the promotions were so ridiculous that local oil men called them sucker tents. Signs offering ten thousand shares of a single well at ten dollars a share were commonplace, and it was not unusual for an individual to be sold a one-five-hundred-thousandth of a one-sixth interest in an oil well.

One leading Los Angeles businessman commented, "There are so many suckers here . . . I know of no place where it is easier to put out a fake security." "People here are especially gullible," a local banker declared. "It is a wonderful, a superb, an unprecedented situation from the promoter's and salesman's standpoint," a leading magazine pointed out. "You must not think only the tourists and hicks from the backwoods buy oil," one financial authority was quoted as saying; even "the more intelligent people are taking a shot at it."

One of the best-known of Signal Hill's promoters was Chauncey C. Julian. As one historian recorded, he was transformed almost overnight from "greasy overalls to tailored clothes, manicured finger nails, spats, a derby hat and cane . . . [driving] a Pierce Arrow." Julian Petroleum Corporation became known throughout the country. "Mail that old check or run in and meet me personally" . . . for "You'll Never Make a Thin Dime Just Lookin' On!" Julian told prospective investors. Within a month of beginning his promotion, he had forty thousand stockholders and eleven million dollars of their money. Julian squandered much of his wealth, and after a running feud with local bankers and stockholders, he was forced to sell his company to S. C. Lewis. Later, Julian left for China, where he commited suicide.

Lewis joined with Jack Bennett to continue Julian's program. Both men arrived in

Los Angeles in 1924, "with all their belonging in a single suitcase." Borrowing $800,000 from local bankers, they purchased the Julian Company and then formed stock pools to sell large blocks of stock below the current rate to financiers who agreed to withhold them from the market for a period of time. To make the scheme appear legitimate, Lewis and Bennett persuaded Motley H. Flint, the executive vice-president of Pacific Southwest Trust and Savings Bank, to involve the bank heavily in the pool.

To many investors it seemed that the pools offered unparalleled profits. Within five months one pool returned a profit of $790,000 on a $1 million investment. In 1926, Lewis's and Bennett's stock reached $35 a share. The bubble burst in April, 1927, when word of the over-issue of Julian Petroleum Corporation stock reached the Los Angeles Stock Market. Although Flint assured investors that "everything is absolutely okay," on April 25 of that year the pooling arrangement folded. The stock price plunged, and the resulting investigation uncovered a huge swindle. Six hundred thousand shares of stock had been authorized, but nearly three million had been sold. Indicted, Lewis and Bennett were acquitted after allegedly bribing Asa Keys, the Los Angeles district attorney. Later, however, they were convicted in connection with another swindle and sent to jail. Flint was shot by an irate investor.

Some promotions were profitable in unexpected ways. In 1906 word spread of an oil strike near Sherman, present-day West Hollywood. Shortly after, Charlie Canfield, Burton Green, and Max Whittier paid $670,000 for the Rodeo de la Aguas, just to the west of Sherman. Although the men were aware that without irrigation the land would produce only beans, they hoped that the production at Sherman would expand to the west. After drilling thirty dusters, the three oil men realized that they had miscalculated. Rather than redevelop the land as a bean farm, Canfield, Green, and Whittier subdivided the property into lots, adopted restrictive covenants prohibiting any petroleum development on the land, and named their new community Beverly Hills.

Soon after the completion of Alamitos No. 1, an extensive town-lot drilling program got under way at Signal Hill. Up to that time Long Beach had been "a quiet little seaside village where retired Iowa farmers pitched horseshoes." Once the discovery well came in, it became a "bustling, hustling city undergoing a feverish development and approaching the 100,000 mark" in population. "The streets are filled with great throngs of people, the traffic is dense, and groups stand on every street corner, discussing oil and real estate."

Almost overnight, derricks appeared among the homes. Some boilers were placed in the streets because there was no room on the actual leases. By July, 1923, 270 wells were drilling, and to one observer the hill had "the appearance of a mammoth porcupine." Some lucky landowners were offered as much as twenty thousand dollars an acre for their property. However, such overdrilling had its price. It was estimated that $37.5 million was wasted by sinking unnecessary wells in the Signal Hill Field. Much of the excess drilling was the result of landowners' insisting on clauses in their leases demanding rapid development of the property.

Workers were desperately needed in the field. Crews worked around the clock because there were not enough men available to split shifts. Once when the Shell Oil Company received several railroad cars of supplies at once, company officers hired anyone available on the spot, until they had the necessary 127 drivers to deliver the equipment to the well sites. With labor in short supply, wages were high. Contract drillers were offered as much as $10.50 per day and roughnecks, $9.00.

Supplies also were in short supply. According to one witness to the Signal Hill boom, there was "'never enough'—never enough material, never enough labor, never enough room." In the early boom there was such a shortage of drilling mud that Shell drivers were sent "miles away to dig it from abandoned sumps."

Newly arrived workers quickly rented all available housing. Latecomers were forced to build shacks in the roads, and traffic just had to maneuver around them. Along with the workers came the oil-field camp followers. Prostitutes abounded, and gamblers found ready marks. Although Prohibition was in effect, liquor was plentiful. The standard method of conducting business with local bootleggers was for a customer to leave his money in a certain eucalyptus tree at night and then return for his whiskey later. One of the region's most famous drinks was made out of persimmons and pineapples and called "one way" because, as one longtime California oil man recalled, "one glass and you were down."

The gamblers designed their own type of building, made out "of one-by-twelve boards with bolts," according to one participant in the Signal Hill boom. These structures had a front and a back room, with a "quick escape back door." "In the front," he continued, "would be pool tables and a tobacco counter; in the back, card tables, and of course, slot machines—five cents, ten cents, twenty-five cents and a dollar. . . . The pool tables were . . . used . . . mostly to bank dice on."

The Old Baker Winery Barn was the best-known gambling den in the Signal Hill–Santa Fe Springs region. The only safe way to visit the establishment after dark was for the "whole crew" to go together, one local resident recalled. "The girls were as bare as the walls. . . . It was like going to a carnival. . . . Once you got in you could see the girls shimmy for nothing . . . [or] shoot craps or play the blackjack table or draw poker or roulette wheels."

The south and east flanks of Signal Hill, which rose to an elevation of 325 feet, were the scene of much of the early development. The Shell Oil Company completed its Horsch No. 1 on October 26, 1921, on the north flank and opened that area to production. Later the west flank was developed also. By April, 1922, only ten months after the completion of the discovery well, Signal Hill boasted 108 wells, with a daily output of 14,000 barrels of crude. The boundaries of the field continued to be extended, and in 1923 productive horizons were uncovered at the 5,000-foot depth. Although in the fall that year the Long Beach Pool was flowing at 259,000 barrels of crude daily, by 1924 most of the wells had stopped flowing freely, and oil men were forced to place them on pumps. By 1925 the field stretched into the Los Cerritos area north of Wardlow Road.

Around 1924, Zeb Dyer conducted one of California's earliest acid bottle tests at the Signal Hill Field. Some of the early wells were plagued with crooked holes. Some curves may have been intentional, as in whipstock drilling, but most were accidental. If the curve was too great, it complicated the setting of casing, and drillers needed some way to determine the amount of angle in a hole. To solve the problem Dyer simply filled a bottle with acid and lowered it into the hole on a line. As the bottle was dropped, any curve in the hole could be measured by the angle of the etching the acid left on the glass.

Between 1927 and 1938 more than three hundred additional deep wells were drilled in the nearby Lovelady and Hilldon areas. By the end of 1938, the Long Beach, or Signal Hill, Field had produced 614.5 million barrels of crude. At that time its total productive area stood at one thousand four hundred acres. This gave the pool an average recovery rate of 440,000 barrels per acre. Such impressive production resulted from the close spacing of the wells and the thick productive horizons. In 1950 Long Beach's cumulative production stood at 750 million barrels of crude. This was an average of more than 500,000 barrels of oil per acre. Such an output made the Signal Hill Field the "richest field in terms of production per acre than the world had ever seen."

Northwest of Long Beach, the Chanslor-Canfield Midway Oil Company opened the Torrance Field to production when it completed the Del Amo No. 1, for a flow of 300 barrels daily, on June 6, 1922. Stretching inland from just south of Hermosa Beach, the pool ran north to south diagonally. Development was slow until several large wells were completed inside the town of Lomita. Afterward the region boomed. Early production peaked in May, 1924, with an output of 72,000 barrels daily from 345 wells. A deeper zone was opened by the Chanslor-Canfield Midway Oil Company on July 22, 1936, when the Del Amo No. 23 was completed at a depth of 4,887 feet. The discovery well flowed at 105 barrels daily and started another rapid drilling effort in the area. Wells with a daily production as high as 700 barrels were reported from this horizon.

By 1941 the Torrance Field had become the largest in the Los Angeles Basin and covered an area seven miles long and a mile wide. By January of that year, approximately 1,200 wells had been drilled in the pool, and total production was estimated at 100 million barrels. In September, 1941, 656 wells were flowing, at a combined 9,277 barrels daily. It was, however, a flush field, and many of the early wells rapidly became unprofitable to operate. Shell, for example, lost four million dollars in the pool.

About fourteen miles south of Los Angeles, one of the most important finds of the Los Angeles Basin, the Dominguez Field, was opened by the Union Oil Company of California in September, 1923. Its discovery well, the Callender 1-A, flowed at 1,193 barrels daily from 4,068 feet. At one time a drilling crew in the pool claimed the world's record for speed drilling when they made 3,250 feet of hole in fifteen days—an average of 217 feet a day.

Most of the acreage in the pool was owned by three landowners, and, as a result, the field was developed slowly with little wasted drilling. It was the site of an unusual accident, however. In the summer of 1925, Shell Oil Company completed its Reyes No. 27. Casing had been set and cement run when enough gas accumulated below the plug to blow the head off the well. "A geyser of sand, water, pine cones, and pieces of petrified wood" spewed into the air. It was two weeks before the well was brought under control. However, just as the runaway was plugged, "with a deafening crash the ground gave way and the boiler, rig, draw-works, platform, and derrick all disappeared into the ground." Only a few feet of the top of the derrick remained visible. The site was never reopened. Instead, it was used as a dumping site for scrap material. Later, during World War II, the pit was cleaned out, and the metal was used for the war effort.

Production in Dominguez peaked in 1925, with an annual output of 4 million barrels of crude. By January, 1941, 314 wells had been drilled the field; but 44 had already been abandoned. As of that date, Dominguez's cumulative production touched 121,800,000 barrels. Average daily production was 20,151 barrels of crude.

West of Los Angeles, near the northwest extremity of the Los Angeles Basin, along a line of folding that ran southeast toward Newport Beach, lay the Inglewood Field. Oil men first began drilling in the Baldwin Hills region in September, 1916, but they were south of the producing horizons. Five dusters were drilled between 1916 and September 28, 1924, when the field was finally opened by Standard Oil Company of California's Los Angeles Investment No. 1-1, which was drilled on what proved to be the extreme southern border of the pool and flowed at 145 barrels a day. Ironically, four of the previous dry holes were within half a mile of the producing horizons, and the other was only a mile away. In late 1924 the Mohawk Oil and Gas Syndicate completed a 250-barrel-per-day producer on the western edge of the pool. Within a year of the original discovery, there were 150 producing wells in the field. Most of the 875 productive acres at Inglewood were held by five companies, so there was little wasteful drilling.

Between the Long Beach and Huntington Beach oil fields, the Seal Beach Field was opened by the Marland Oil Company, which later became the Continental Oil Company, in August, 1926. Its discovery well, the Bixby No. 2, flowed at 1,240 barrels daily. Rapidly developed, especially in the town-lot area of Alamitos Heights, the pool became one of the most prolific producers along the Newport Beach–Inglewood line of folding. Production in the Seal Beach Field peaked in June, 1927, at 75,000 barrels daily.

Southeast of Inglewood and along the same fault zone, which stretched from Culver City to Huntington Beach—the Inglewood Fault Zone—the Potrero Field was uncovered. It was eventually expanded to include the Cypress Pool, the Townsite Pool, the Potrero Area, and the East Townsite Block. Total production for the four fields was more than 3 million barrels of oil as of June, 1941.

Several other smaller fields were located in the Los Angeles Basin in the 1920s and 1930s. Just to the southeast of El Segundo, the San Clemente Oil Company opened the Lawndale Pool on July 3, 1928, inside the community of Lawndale. In August, 1929, in the region along Santa Monica Bay to the west of Los Angeles, near Venice and the neaby tidal marshlands of Ballona Creek, the Venice–Del Rey Field was opened. Four miles south of the Playa Del Rey Field, the Republic Petroleum Company opened the El Segundo Pool on August 30, 1935.

In 1936 the last of the major finds in the Los Angeles Basin was made at Wilmington, along the northwest edge of San Pedro Bay, between the center of the town of Wilmington and the Los Angeles County Flood Control Channel in Long Beach. The Wilmington Field was opened to production on December 6, 1936, by the General Petroleum Company, whose Terminal No. 1 flowed at 1,389 barrels daily. By November, 1940, Wilmington's 914 wells were flowing at 81,000 barrels daily.

Much of the oil at Wilmington could be found at a rather shallow depth, making natural seeps common. After the strike, some of this shallow crude was tapped as an extra source of income by area residents. The boom in shallow wells started when Cristobal Salcido was digging a twenty-foot-deep cesspool in the back yard of his home in Wilmington. Salcido was surprised when the hole suddenly began to fill with light oil. The crude had such a high gravity that it could be used for motor fuel, and Salcido began collecting the oil in barrels and then selling it to local motorists. As word spread, many of his neighbors dug similar pits before local authorities could stop the practice. Later the city council agreed to permit the shallow-pit wells to resume production if the owner, among other requirements, posted a one-hundred-dollar bond, kept a fire extinguisher close by, and refilled the pits when they no longer produced. Also, all pumping had to be done by hand. Even so, within a short time the country became "dotted with fields of midget wells."

Wilmington became a major refining center for the Los Angeles Basin. In March, 1919, C. W. Dubbs had patented a "clean circulation" process to refine crude. One of the first full-scale Dubbs process plants constructed on the West Coast was built at Wilmington by the Shell Oil Company.

The construction of heavy industry along the ocean front at Wilmington helped create one of the most difficult problems for California oil men: sinking land. In the late 1940s, it was suddenly discovered that the central beach area of the Wilmington Field was sinking. The lowering of the seacoast was so pronounced that by 1950 it was subsiding at the rate of two inches per month. To offset the sinkage, local construction engineers were forced to tunnel deep underground to shore up foundations. Even so, many streets, sewers, pipelines, and underground electrical cables broke under the stress.

An investigation of the phenomenon disclosed that, although Wilmington's oil reservoirs were exceedingly thick, the overlying shale was not. As a result, as the oil and natural gas were removed, at the rate of 45 million barrels of oil and 25 million cubic feet of natural gas annually, water was prevented from filling the void by underground rock barriers. These conditions, plus the fact that the field's underground dips were relatively flat, allowed the overlying earth to begin stinking as the oil was pumped to the surface.

To solve the problem, oil men began replacing the withdrawn oil and gas with water by artificial methods. Local and state legislation was passed to provide for the injection of pressurized water. Several water-injection plants were constructed, and by

1968 it appeared that the subsidence had been halted, and some of the land was being uplifted.

By 1949, Wilmington had become the seventh greatest producing oil field in America, and in that year it ranked as the second best producing field in the state, with an output of 48,317,000 barrels of crude. Cumulative production at the close of the first half of the twentieth century was 460,880,000 barrels of oil. In 1964 Wilmington became the first field in California, and the second pool in the nation, to produce 1 billion barrels of crude.

Although the strike at Wilmington marked the end of the Los Angeles Basin boom, it helped California maintain its ranking, second only to Texas, among the nation's oil-producing states. In 1923 the production of Signal Hill, Santa Fe Springs, and Huntington Beach, combined with the smaller pools of the Los Angeles Basin, accounted for 20 percent of the world's total production of crude. In the fifteen months preceding December, 1923, "approximately 80,000,000 barrels of crude were shipped from California through the Panama Canal to Gulf, Atlantic, and European ports." Such a "black flood . . . placed California in the premier place among producers."

The Roaring Twenties ushered in the American automobile age and opened an immense new market for California oil. To tap the new market many California oil companies opened their own chains of "service stations," which catered to the motoring public. This Union Oil 76 station at the corner of Adams and Grand streets in Los Angeles reflected the mission style of architecture so prevalent in the state. *Pacific Oil World*

As Californians became more and more enamored of the automobile, service stations proliferated. To accommodate the drastic rise in the number of service stations, the California Oil Company adopted specialized gasoline-delivery trucks to carry their products throughout the state. This Red Crown Gasoline truck has just delivered a load of "new winter" gasoline to the combination service station–grocery store in the background. *Standard Oil Company of California*

By the beginning of the 1920s, wells in the Los Angeles Field had penetrated into the heart of the city and were common sights in residential areas. However, local ordinances had placed some regulations on the oil companies. For example, the side of the derrick nearest the house has been covered with sheet-iron so that, in case of a wild well, the home would be protected from the blowing oil. *American Petroleum Institute*

At Huntington Beach, just to the south of Los Angeles, the derricks and bathers shared the beach. By this time technology had developed to the point that many of the wells directly behind the beach had slanted their holes, so that they were pumping either from beneath the bathers or from farther offshore. *Standard Oil Company of California*

This view of Huntington Beach shows vividly how close the wells were to each other. Because much of the property had been subdivided into twenty-by-twenty-five-foot lots, leases became a major headache for developers of the pool. *PennWell Publishing Company*

An aerial view of the Huntington Beach Field, where town-lot leases created a maze of well sites, with as many as three wells, such as those in the right center, lined up one after the other. The storage tanks in the right center have been surrounded by earthen dikes to prevent any spilled oil from spreading. *Security Pacific National Bank Collection, Los Angeles Public Library*

Huntington Beach was also called "Bedlam Beach" by area oil men because of the seemingly impossible situation they often found themselves in when drilling in the pool's small leases. It was not uncommon for two wells to be drilled from a single twenty-five-foot lease. This well of the Southern California Petroleum Company is wedged between a house and an apartment, with the tank battery in the house's backyard. *PennWell Publishing Company*

Boiler Row at a Huntington Beach refinery. The boilers were used to supply power to the refining facility. Note that some are brick-walled, while others have their steel sides exposed. *Atlantic Richfield Company*

Potential investors in the Los Angeles oil boom—and probably sightseers, as well—stand before a circus tent while waiting for the free lunch offered by one of the numerous promoters selling shares in wells. *PennWell Publishing Company*

Left: Sightseers and would-be investors crowd around a well at Compton in 1926. Standing on the platform, the promoters of the operation are getting ready to give their "pitch" to encourage investment in the well. *Security Pacific National Bank Collection, Los Angeles Public Library. Right:* Shell Oil Company's Alamitos No. 1, the discovery well of the Signal Hill, or Long Beach, Field. Opened in 1921, Signal Hill was uncovered just as the oil industry was switching from wooden to steel derricks. While the discovery well was drilled by a wooden derrick, just to the right is a more modern steel structure. *American Petroleum Institute*

Shell Oil Company's Andrews No. 3 gushing oil at Signal Hill in 1923. At this time the actual hill, clearly visible in the background, had not yet been completely covered with derricks. *American Petroleum Institute*

Next to Signal Hill was another small rise, shown here in the foreground. Because a large storage tank was constructed on the smaller hill, it was appropriately named Reservoir Hill. *Atlantic Richfield Company*

A 1932 aerial panorama of Signal Hill taken from above the hill's crest. Notice the multitude of small leases on the hill's flanks as a result of the area's subdivision into town lots. *Atlantic Richfield Company*

This view of Signal Hill was taken two decades after the initial discovery. By this time a forest of derricks literally covered the hill. *American Petroleum Institute*

In this photograph of Signal Hill, the cemetery in the foreground is the only piece of property not containing a drilling rig. Notice the dike built around the cemetery to prevent any spilled crude from desecrating the graves. PennWell Publishing Company

A night view of Signal Hill. So many gas flares are visible that some wells did not need any other lighting to operate at night. *Security Pacific National Bank Collection, Los Angeles Public Library*

The Richfield Oil Company's complex at Signal Hill. To streamline its operations in the field, the company created a centralized distribution center for oil-well drilling equipment. Note the directional arrow painted on top of the building. It was not uncommon in the early twentieth century to paint such directional indicators to help pilots. *Atlantic Richfield Company*

Left: The home of Richfield Oil Company's drilling superintendent in the Signal Hill Field. Even the well-kept grounds of the firm's headquarters provided a location for several wells. The well immediately to the left of the building, with pipe stacked inside the derrick, was for a short time in May of 1928 the deepest oil well in the world. *Henry E. Huntington Library and Art Gallery. Right:* Because of the subdivision of the area into town lots, Signal Hill was overcrowded by derricks, each built on a single lease. In this photograph the legs of neighboring derricks seem almost to be interlocked. The tree trunk in the center foreground provides mute testimony to the need for poles to be used as the cellars of wells. *American Petroleum Institute*

High concentrations of natural gas posed a constant danger to oil men developing the Signal Hill Field and made blowouts and fires common occurrences. In this photograph the man and two women in the center seem unconcerned about the burning well in the background. However, in the extreme center background quite a crowd has gathered on a small rise to watch the fire. *Standard Oil Company of California*

Signal Hill was one of the most extensively developed fields in the world. Notice the tremendous amount of drilling activity still under way in this photograph, indicated by the steam escaping from the boilers, even though at least eighty-four derricks are already in place. One lone palm tree still stands along the road in the center foreground, amid the forest of derricks. *PennWell Publishing Company*

Although there were numerous small leases at Signal Hill, this one in 1937 claimed the distinction of being the "world's smallest producing lease." The entire lease was enclosed by the fence and barely had room for the derrick and a storage tank. *PennWell Publishing Company*

In 1923 the Petroleum Midway Oil Company installed the world's first electric motors driving centrifugal pumps in its Watson Station at Signal Hill. They were 200-horsepower squirrel-cage induction motors manufactured by Westinghouse. *PennWell Publishing Company*

In the early 1950s there was a revival of drilling activity on Signal Hill, as deep producing horizons were uncovered by the Texas Company. This is Hancock Oil Company's Signal No. 52, which flowed at 1,922 barrels daily. *PennWell Publishing Company*

Left: Supply Row at Long Beach was so named because the street was lined with oil-well supply houses. Ironically, when drilling at Signal Hill was renewed in the 1950s, much of the deeper production was found along Supply Row, and within a short time derricks were visible in nearly every available space between the supply houses. The advance in technology allowed many of the later wells to be drilled with portable steel drilling rigs rather than the older wooden or permanent steel structures. *PennWell Publishing Company. Right:* A part of the huge Richfield Oil Company complex at Long Beach Harbor. The facility is surrounded by drilling rigs that extend up to the water's edge. The air vents and concrete covers in the photograph enclose a portion of the massive pipeline complex within the plant. *PennWell Publishing Company*

A gusher in the middle of a Japanese vegetable garden at Torrance. Obviously the spewing oil ruined the surrounding plants; however, the landowner undoubtedly was more than compensated by the royalty received from the well. *Security Pacific National Bank Collection, Los Angeles Public Library*

Richfield Oil Company's tanker *Richfield* outbound from Long Beach Harbor for the Oakland–San Francisco Bay area. The *Richfield* was originally built in 1913 for the coastal tanker trade and was used until it went aground and broke up on the rocks at Point Reyes, California, on May 8, 1930. Because of its location near readily available shipping facilities at Long Beach, Signal Hill became a major supplier of crude for shipment overseas or to other markets on the West Coast. *Atlantic Richfield Company*

A U.S. Navy tanker of the Rapidan class onloading crude at Long Beach. Tankers of this class could carry 11,145 tons of oil fuel. A part of the Navy's fleet train, the tanker would refuel combat ships by passing large hoses from its tanks to the fuel tanks of the other ship as they steamed side by side at a slow speed. Such refueling techniques, mastered by the Navy after it switched from coal to oil, allowed the fleet to remain at sea for an extended period of time. *PennWell Publishing Company*

A tank battery and pumping motors in the Dominguez Field, just south of Los Angeles. Note the multitude of pipelines running into the facility on the right. Lines from thirty-nine wells empty into this single tank battery. *PennWell Publishing Company*

The Inglewood Field in August, 1925. Unlike those in many of the other fields of the Los Angeles Basin, the wells at Inglewood were regularly spaced since most of the region's 875 producing acres were owned by only five oil companies. This is one of the few examples of an orderly development rather than a mad scramble for town lots in the area. *American Petroleum Institute*

An aerial view of the Inglewood Field in June, 1930. Although located within the community, Inglewood was not a typical town-lot field. Instead of wasteful drilling of unnecessary wells, Inglewood was notable for its orderly spacing of the derricks, as shown here, upper left. *Atlantic Richfield Company*

The Standard Oil Company of California's El Segundo Refinery. Located in the midst of the Los Angeles Basin, El Segundo became a major refining center. The teams of mules in the photograph are excavating another huge oil storage tank at the refinery to enable it to store the tremendous production from the nearby wells. The size of the storage tanks can be easily judged by comparing the one in the background with the railroad tank cars visible along its rim. *Standard Oil Company of California*

The Venice–Del Rey Field along the tidal marshlands of Ballona Creek near Venice in 1936. Many of the wells sharing the beach with the houses have slanted their holes offshore. Notice that it is extremely difficult to find a single beachfront lot that does not contain a drilling rig. However, the farther inland the wells are, the more widely spaced they are. *Standard Oil Company of California*

Bixby Slough in the Wilmington Field. Note the oil-covered water in the right foreground. Oil men were first attracted to the region by such oil seeps. In the left foreground is a portion of a refinery and tank farm. The Wilmington area became a major refining center for the California petroleum industry. Most of the drilling activity is concentrated in the flat low-lying region in the back of the photograph. *Standard Oil Company of California*

This photograph clearly shows the effect of subsidence on oil-well pumping operations in the Wilmington Field. Notice the wellheads on the right and in the center of the photograph. Extensions of the oil-well casing, the white, vertical posts, will be below ground when the land is filled in there. *PennWell Publishing Company*

In this photograph the tanks in the background lie eighteen feet below the elevated wellheads in the foreground, as a result of subsidence in the Wilmington Field. *PennWell Publishing Company*

Although the removal of underground oil was to blame for much of the subsidence at Wilmington, the location of such heavy industries as the plant visible in the right foreground also contributed to the sinking. Notice the tankers tied up in the second loop of the slough, waiting to load crude. Mixed with the modern ships are several older sailing ships, which, no longer used to transport crude, had been relegated to serving as storage vessels. *Standard Oil Company of California*

Coalinga and Northern California

As the Roaring Twenties were ending, California oil men turned their attention northward to the Sacramento Valley, the northern San Joaquin Valley, and the Northern Coast Ranges. Although most of California's rich mineral deposits are found below the San Francisco Bay area, the region from San Joaquin County northward to Humboldt County produced some rich natural gas discoveries and a few minor oil finds. Most of these petroleum deposits were not tapped until the 1930s.

In the foothills of the Diablo Mountains in Fresno County, the huge Coalinga complex was uncovered at the beginning of the twentieth century. Stretching north from the town of Coalinga, the myriad of fields includes the Coalinga West, Coalinga East, East Coalinga Extension–Eocene (Coalinga Nose), Oil City, and Northeast Coalinga Eocene pools. The region had a history of fossil-fuel production. The town of Coalinga was named for the nearby deposits of low-grade coal, which area residents burned. For a while in the late nineteenth century, before the switch from coal to oil by the railroad, the community was the center of a coal boom.

Earliest oil development in the area started in 1898, when the Oil City Field was opened to production. The Blue Goose, completed that year by California Oilfields, Ltd., was one of Coalinga's most famous early-day gushers. Reportedly the largest well in the pool, it initially flowed at seven hundred barrels daily; however, by 1900 production had dropped to an average of fifteen to twenty barrels daily per well. Oil City's output continued to decline, until by 1936 the average daily well production was a modest four to six barrels.

In 1901, three years after the Oil City strike, the Coalinga West Pool was opened. Separated from the rest of the Coalinga complex by a syncline, Coalinga West produced some impressive gushers; one of the most famous was the Silvertip, drilled by the Silvertip Oil Company. Located about a mile and a half outside the town of Coalinga, the Silvertip was completed in September, 1909, as a producer of 10,000 to 20,000 barrels per day. It blew in unexpectedly while the crew was bailing, and the roar from the escaping natural gas was so great that it reminded onlookers of "locomotives blowing off steam." Workmen rushed to build dikes around the drilling site to collect the spewing crude. Oil was thrown so high into the air that it was carried half a mile by the wind before falling to the ground. In one seventy-two-hour period, 36,000 barrels of crude were trapped in nearby sumps and gullies. Within a month the Silvertip produced 102,000 barrels of crude.

In 1904, the Coalinga Oil Transportation Company, a subsidiary of the Associated Oil Company, built another six-inch line connecting Coalinga with the seaport of Monterey, 110 miles to the northwest. In addition, an eight-inch branch line was built to connect the Coalinga wells with Mendota, where it joined the Standard Oil Company's "main Kern River–Point Richmond line." With the completion of these pipelines, the boom at Coalinga really got under way.

The town of Coalinga boomed overnight. One of the community's largest business establishments was the Coalinga Lumber Company, which opened in November, 1907. It was the major supplier of timbers for building rigs for the pool and generally kept 750,000 feet of lumber in stock. Within a short time of the opening of the nearby oil field, a host of "tar-paper shacks, which would be glorified by the name 'bunkhouse'" had been built in the community.

Just getting to the new strike was a problem for oil men. The town had no railroad depot. The nearest stop was twenty miles away, at Huron. Once off the train, workers started across the desert on their own. According to one old-timer, "You'd start out over those deserts and when the ruts got too deep, you'd just move over." Within a short time the road was half a mile wide in some spots. Oil men and investors poured into the community. New arrivals were greeted by the local band and served banquets. To attract attention and draw crowds, the Coalinga Athletic Club once raised ten thousand dollars to promote a championship boxing match between Jim Jeffries and Jack Johnson.

Front Street, better known as Whiskey Row by local inhabitants, was the center of Coalinga's roughneck district. Such businesses as the Portola Bar and Grill, the Axtell, and the Palace lined the street. Their main attraction was bad food and plentiful liquor. The streets were nothing but mud. Along their edges, board sidewalks, sometimes as high as three feet off the ground, offered sanctuary to pedestrians. It was almost impossible to make any headway against the crowd. Strollers were often forced off the sidewalks into the muck. Few physical comforts were available. One sign in the community advertised baths. The price was twenty-five cents for first use of a tub of water, fifteen cents for "seconds."

Gambling was rampant. One visitor to the town recalled that "every establishment in the block [Whiskey Row] had several tables of poker and blackjack constantly in progress." "Stacks of twenty, ten, and five dollar gold pieces [were] piled up like poker chips, and it wasn't uncommon to see awhole stack of these gold pieces shoved to the middle of the table on a single bet."

In 1901, the year following the opening of Coalinga West, the first well was drilled in the Coalinga East Pool. For several years William M. Graham had entertained the idea that East Coalinga contained oil. Early in 1901, he convinced the firm of Balfour, Williamson & Company, of London, England, to finance a wildcat in the region. They then purchased a large share of California Oilfields, Ltd., also a British company, when it was formed.

Between 1902 and 1903, most of the region's development was undertaken by California Oilfields, with some of the wells initially flowing at between 3,000 and 4,000 barrels daily. One well was reportedly completed as a 7,000-barrel-per-day producer. As late as 1937 some of the earlier wells were still flowing at about 300 barrels daily; however, the average was between 50 and 100 barrels.

A major developer in the state, California Oilfields, was purchased by Shell Oil Company in 1913. At that time the properties of California Oilfields accounted for 4.5 percent of California's total production. Its acquisition by Shell gave the Dutch firm its first sizable production on the Pacific slope.

In an attempt to offset the more primitive living conditions available in Coalinga's more notorious boom towns, California Oilfields, backed by Balfour and at Graham's insistence, undertook the construction of a "model oil town." Originally, the community was named Balfour, but later the name was changed to Oilfields. By 1908 it was in operation. A specially designed building, "the Bungalow," dominated the community. It was a large "H-shaped, one-story building on the crest of a small hill overlooking the rest of the camp." Designed for "unmarried white-collor, or 'staff,' employees," the central portion of the Bungalow contained "luxurious living, library, and dining rooms," while the wings were divided into private living quarters, each with an adjoining bath. It was staffed with "a Scotch gardener and Japanese houseboys."

Surrounding the Bungalow was the town of Oilfields, a "complete, self-contained, almost wholely self-sufficient community." It included several tool shops, a tin shop, a tank shop, a wood-planing plant, and a brickyard. These facilities could fabricate practically all of the variety of tools and equipment needed to maintain the company's operations. Oilfields also boasted a water-condensing plant, an ice cream plant, a steam plant, an electric generating plant, and a multitude of warehouses and storehouses.

To provide public transportation, "Big Betsy," a large steam tractor, pulled a train of trackless cars between Oilfields and Coalinga. A nearby farming operation, also run by the company, supplied "vegetables, milk, butter, fresh eggs, poultry, hams, pork, and beef." A small cooperative store sold basic necessities. The town even had its own post office. The population of the community was usually around three hundred men. To provide entertainment during non-working hours, company-sponsored baseball, basketball, soccer, and tennis teams took the field, and amateur theatricals, glee clubs, and weekly motion pictures were regular events. In addition, a company golf course and swimming pool were available.

The introduction of rotary rigs in California reportedly took place in 1902 at Coalinga. W. K. Oil Company used one to sink a hole to the 2,400-foot level only to discover that the hole was so crooked the crew could not set casing. To complete the well, a cable tool rig was moved in, the hole redrilled to eliminate the crooked sections, and casing set. The failure postponed the widespread adoption of the rotary rig for nearly a decade.

With the increased output of the Los Angeles Basin in the early 1920s, many California producers were hard-pressed to provide adequate storage space for their crude. In an effort to ease the oil glut and to save Coalinga's reserves for the future, late in 1922 Shell Oil Company of California closed 305 of its Coalinga wells, which had accounted for a daily output of 16,000 barrels. It was not until 1932 that Shell began gradually to reopen Coalinga's production.

On June 26, 1938, the East Coalinga Extension was opened, to the south of Coalinga in the Kettleman North Dome Field. The discovery well, the Gatchell No. 1, was drilled by Petroleum Securities and was completed as a 10,000-barrel-per-day producer. Later, on April 24, 1939, the Amerada Petroleum Corporation located the Eocene Pool about a mile to the east of the East Coalinga Extension, when it completed its S.P.L. No. 7-17. These two pools eventually were merged into a single field but remained divided into the Amerada and Gatchell areas. By June 1, 1941, 120 producing wells had been drilled in the Gatchell Area. Development of the Amerada Area was undertaken almost exclusively by the Amerada Petroleum Corporation, and as of July 1 of the same year, 64 producing wells had been completed in it. The productive area of the combined region totaled about thirty-seven hundred acres—twenty-five hundred in the Gatchell Area and twelve hundred in the Amerada Area.

Northern California was also rich in natural gas. The area had a history of natural gas seeps, and as early as 1854 a natural gas well was drilled near the Stockton courthouse. By 1858 it had reached a depth of 1,003 feet and its gas was being sold commercially. Many water wells in the region were also found to contain natural gas deposits. Much of the gas was used locally in manufacturing and for lighting and heating homes. Later, in 1864, the natural gas spewing from an artesian well at Stockton was tapped and then piped into town to be used for lighting and heating. In 1885 the Standard Gaslight and Fuel Company was formed at Merced to develop the natural gas deposits of the San Joaquin Valley. The following year the California Well Company was also organized at Stockton for the same purpose. Several wells were drilled by the firms. One, the Crown Mills well drilled in 1886, flowed at 18,000 cubic feet per hour.

In 1891 natural gas was located in water wells near Sacramento. It too was put to local use. In Humboldt County, near Briceland, a well was drilled that began supplying the community with natural gas in May, 1894. In 1901, the Rochester Oil Company drilled a well in Solano County to supply Suisun and Fairfield with gas. Natural gas seeps were well-known in the region along Bear Creek in Colusa County, to the northwest of Sacramento. In 1864, Dexter Cook, while digging a water well on the south flank of Sutter Buttes, about twelve miles west and a little north of Marysville in Sutter County, had uncovered a natural gas seep. Hoping that the natural gas indicated the presence of a vein of coal, Cook started a shaft near the seep in February, 1864, but he encountered only more natural gas. While digging in the mine, he ignited the escaping fumes; although unhurt by the fire, Cook returned to digging water wells.

Unknowingly, Cook had located the Marysville Gas Field, also known as the Sutter Buttes Gas Field. Commercial development of the region was started in 1927 by the Sutter Buttes Oil Company. The area was geologically and structurally unique in California. The center of the pool, which was nearly circular, was the mile-wide crater of an extinct volcano, filled with debris and surrounded by three roughly concentric topographic zones. However, the Marysville Gas Field was not examined seriously until 1932, when Walter Stalder, with the financial backing of Buttes Oilfield, Inc., spudded in his Buttes No. 1 and Buttes No. 2 in the section west of the earlier holes. However, only a negligible amount of natural gas was found.

Even farther north, in central Humboldt County, the Texas Company carried on extensive exploration north of Loleta. As early as 1861 a well had been drilled on the Davis Ranch to tap a natural oil seep. No commercial oil was found, but substantial quantities of natural gas were located. Wildcatting in Humboldt County, however, was hampered by inadequate financing.

The first commercial gas field uncovered in the northern part of California was found in 1932, near Tracy in San Joaquin County, about twenty miles southwest of Stockton. After completing a thorough seismographic exploration of the area, the Amerada Petroleum Corporation located the Tracy Gas Field in 1935. Tracy's production peaked the next year at 3,012,083,000 cubic feet. By September, 1942, six gas wells had been completed in the field, four of which were still producing.

One of northern California's most unusual gas fields was the McDonald Island Gas Field, located about ten miles northwest of Stockton. An extensive reflection-seismographic survey, started in 1935 by Geophysical Service, Inc., under the direction of Standard Oil Company of California, led to its discovery.

In March, 1936, the McDonald Island Farms No. 1 was spudded in. When tested, the well showed 8 million cubic feet per day at 4,200 feet, and when it was completed in June, 1936, it flowed at 26 million cubic feet of dry gas daily. Additional drilling uncovered a pool bordered on the north by the Columbia Cut, which runs east and west to join the San Joaquin and Middle rivers, and on the south by the Empire and Turner cuts, which run east from the Middle River to Whiskey Slough and the San Joaquin River.

Also, in June, 1936, the Amerada Petroleum Corporation opened to production the Rio Vista Gas Field, which stretches along both sides of the Sacramento River in Solano and Sacramento counties. The discovery well, the Emigh No. 1, flowed at 81 million cubic feet of natural gas per day. Afterward, the entire area around the town of Rio Vista was developed rapidly.

The Rio Vista Gas Field was somewhat difficult to develop because of the many branches, channels, and sloughs of the Sacramento and San Joaquin rivers, which meander through the area. Most of the waterways were navigable, and thus the minerals were owned by the State of California. To tap these pools, oil men had to resort to slant drilling from the shores. Nonetheless, the Amerada Petroleum Corporation, Standard Oil Company of California, Bishop Oil Company, Superior Oil Company, and a few other firms had completed thirty wells in the field by September 15, 1940; twenty-seven were gassers.

In 1937, the Fairfield Knolls Gas Field was uncovered by the Standard Oil Company about three miles north of Putah Creek, halfway between the towns of Davis and Winters in Yolo County. Although the discovery well, the Hooper No. 1, and Standard's second well in the field both found plentiful gas, with no available means of marketing it, development lagged.

After examining seismographic reports of the region, the Ohio Oil Company spudded in its Willard No. 1 on November 17, 1937, about six miles northeast of Willows in the Sacramento Valley. On January 8, 1938, the well blew in out of control, and within a day a large crater had swallowed the derrick and machinery. Before the well died three weeks later, it was estimated to have a flow of 60 million cubic feet daily. A second well, the Willard No. 1-A, flowed at 5,236,000 cubic feet a day from between 2,237 and 2,245 feet.

Another natural gas strike was made in April, 1938, when the Richfield Oil Corporation spudded in the Potrero No. 1, a little southeast of Fairfield and Suisun City in the Potrero Hills. Previously, the only drilling in the region had been by the Honolulu Oil Corporation in 1921. Originally Richfield's hole was pushed to 5,334 feet, but it was plugged back to 3,265 feet. When it was completed in January, 1940, the Potrero No. 1 flowed at 5 million cubic feet per day.

Throughout the state's long petroleum history, the plentiful reserves of natural gas had been utilized by nearby communities, and in 1907 the Santa Maria Gas Company had begun the first delivery of natural gas directly to customers. By the late 1930s, most of northern California's deposits of natural gas had been tapped. These northern fields, together with the tremendous natural gas production in the southern part of the state, by 1940 ranked California first among the states in the number of customers served by natural gas.

Left: A Cryer & Halbouty well in front of the Oil City Baptist Church. At Oil City the wells penetrated the heart of the residential community. Other abandoned derricks are clearly visible in the background. *PennWell Publishing Company. Right:* Coalinga East, about 1905. On the left are bunkhouses for workers, and to their right are shops to service the wells and other equipment. The pipeline running from the well in the center to the storage tank on the right side of the arroyo is supported across the depression by a makeshift trestle. *American Petroleum Institute*

Oil scouts at Coalinga in the early 1900s. The Silvertip well in the Coalinga West Pool attracted the attention of the California oil industry and triggered a rush to the area. While the two men in the center discuss, probably, potential drilling sites, their drivers wait beside their open touring cars. The man on the right is wearing a driving coat, helmet, and goggles as protection against the elements. *American Petroleum Institute*

A pipeline gang screwing together sections of the six-inch line connecting Coalinga with the Pacific coast. The men on the left levered the pipe into position with wooden braces; then, as the man in the background hammered on the pipe to ensure a tight fit, the men on the right screwed the pipe together with the large tongs. Four men were used on each tong to get the pipe as tight as possible. *Standard Oil Company of California*

A pipe-screwing machine in operation over a creek bed near Los Angeles in 1913. The machine replaced an entire crew of workers, by mechanically connecting the length of screw pipe. The pipe was raised by the bar extending from the front of the machine and held in place while it was screwed together. *PennWell Publishing Company*

Shell Oil Company's huge shop complex in the village of Oilfields, in the Coalinga Field, as it appeared in 1919. Because of the region's isolation, Shell's operations in the pool were almost entirely self-supporting. By this time, the community boasted oiled streets, trees, and more permanent buildings. *American Petroleum Institute*

A view of the immense refinery complex that developed in Coalinga, which made the community a major oil center. The storage tanks on the left have been separated from the actual refining part of the facility by a large earthen moat in order to prevent any fire in the storage tanks from spreading to the refining equipment. *American Petroleum Institute*

Looking past company-owned houses at Oilfields toward the drilling operations on the outskirts of the camp. Shell eventually drilled nearly 300 wells in the area. The housing shown here was available to high-level employees, who brought their families with them to the field. *American Petroleum Institute*

The community of Oilfields was located in the heart of Shell Oil Company's properties in the Coalinga Field in 1920. The town was established by California Oilfields, Ltd., an oil company that was acquired by Shell in 1913. To attract workers to the area, Shell's camp offered large, comfortable housing, surrounded on all levels by broad porches. In the right foreground are a baseball stadium and recreation hall. Trees have been planted throughout the complex to provide shade and an attractive landscape, and street lights have been installed along the streets. *American Petroleum Institute*

"Big Betsy," the large steam tractor that pulled a train of trackless cars between the Shell Oil Company camp at Oilfields and the town of Coalinga. The boiler was supported by the two large steel wheels in the rear, and steering was accomplished by the single smaller wheel in front. *Standard Oil Company of California*

Left: Cable tool rig at Coalinga. Perching beside the rig, the driller controlled the tension on the cable in the hole by the small wheel in his right hand. It was all a matter of guess, and it took a special feel for the equipment to know when to let out more cable or when there was too much slack in the rope. *Standard Oil Company of California. Right:* Cable tool rig in operation at Coalinga. A man is standing beneath the walking beam, which raises and lowers the cable in the hole. Immediately behind the worker is the ladder that gives him access to the pivot of the walking beam on top of the sampson post. Periodically the pivot had to be greased and serviced. *Standard Oil Company of California*

Workers posing before a tool shed at Coalinga. Enclosing boilers within sheet-iron buildings protected the equipment but made the insides unbearably hot in the summer. Note the whistle between the two smokestacks. It was used as an emergency signal should a fire or blowout occur. *Standard Oil Company of California*

Some of the mounted workers at Coalinga used to patrol the well sites to keep livestock out. The heavy wool chaps the men are wearing on their legs were necessary when they were brush popping (chasing cattle through the underbrush) to protect the riders from the sharp limbs. *Standard Oil Company of California*

Typical well site at Coalinga in 1911. The wooden box in the foreground was used to clean oil-soaked clothes. The clothes were placed inside the box, and then steam was forced through the clothes and out the holes to remove the stains. *Standard Oil Company of California*

One of the drilling crews of Standard Oil Company of California posing on the drilling floor of a cable tool rig at Coalinga. From the looks of the oil-covered men and equipment, the well has just been completed. *Standard Oil Company of California*

The aftermath of a boiler explosion at Coalinga in 1912. The danger of a boiler's exploding was always present on cable tool rigs, and, as the photograph attests, the damage when one did was usually extensive. Because some of the area was used as rangeland, fences like the one in the background were built to keep livestock away from drilling sites. *Standard Oil Company of California*

A panorama of the Coalinga Field taken in 1919. The boilers are protected by brick walls, which, if the boiler did explode, would do great damage to equipment and workers. Notice the "black dog," a powerful lamp suspended from the derrick to illuminate the drilling site after dark. *American Petroleum Institute*

Left: The Associated Oil Company's Cypress No. 1, discovery well of the Potrero Field. *PennWell Publishing Company. Right:* Standard Oil Company of California's first well at Coalinga Nose. Nothing remains but the concrete foundation of the drilling rig and the Christmas tree over the well head. *Standard Oil Company of California*

A panorama of the Coalinga Nose Pool after a fire had swept a part of the field. The fire burned an entire square mile of land before the surrounding section-line roads halted it. Notice that many of the roads outside the burned section have been covered with oil to keep down dust. *Standard Oil Company of California*

Standard Oil Company's Sacramento office. Horatio T. Harper, standing beside the cases of Pearl Kerosene Oil, later became a vice-president of Standard Oil of California. Sacramento was provided natural gas from many of the nearby gas fields in northern California. *Standard Oil Company of California*

The remains of the Pacific Coast Oil and Gas Company's offices in San Francisco after the 1906 earthquake and fire. Because of the plentiful supply of natural gas from nearby fields, San Francisco used a great deal of gas for lighting and heating. When the great earthquake struck, it ruptured many of the community's gas mains. The escaping gas was ignited, and the resulting fires caused as much if not more damage than the earthquake. *Standard Oil Company of California*

Much of early-day San Francisco was served by horse-drawn tank wagons selling petroleum products. This is John Krayenbuhl's wagon, from which he served the Haight-Ashbury District of the city in 1915. The 750-gallon tank was divided into three compartments, which held Pearl Oil Kerosene, Red Crown Gasoline, and Zerolene Motor Oil, products of the Standard Oil Company of California. *American Petroleum Institute*

By 1915, the early delivery wagons were being replaced by service stations built at numerous locations throughout San Francisco. This service station of Shell Oil Company of California boasted the latest in gasoline pumps. When an automobile pulled up beside the pump, the hand-operated crank on top of the pump was turned until the desired amount of gasoline was pumped into the glass cylinder. Then the attached hose was placed in the fuel tank, the valve where the hose connected to the pipe was opened, and the gasoline in the cylinder was emptied into the tank. The delivery truck is a chain-driven, Kelly-Springfield truck equipped with a 1,260-gallon tank. *American Petroleum Institute*

A more modern Shell Oil Company service station at Eureka, California, in 1928. By this time the hand-driven pump had been replaced by an electric pump. The large dial on the front of the pumping unit indicated the amount of gasoline sold. *American Petroleum Institute*

In 1915 San Francisco hosted the Panama-Pacific International Exposition, and Shell Oil Company of California was granted the parking concession for the World's Fair. In addition, Shell constructed two red-and-yellow service stations at the exposition's gates. The portable, wheel-mounted pump units in front of the station were used to fill the automobiles as they were parked. Here a "barker" beside the station calls to motorists to park their vehicles in the officially approved parking station. *American Petroleum Institute*

Across San Francisco Bay from San Francisco, and to the north of Oakland, the Standard Oil Company of California constructed a large refining complex shortly after the turn of the twentieth century. Although several permanent buildings have been completed in this 1902 photograph, building is still under way. On the right the progressive stages of work on several large steel storage tanks can be seen. Work on the one at the far right has just started; next to it is another tank, whose sides are being completed; above that, still another tank is receiving the finishing touches; and last in line is a completed tank. *Standard Oil Company of California*

The "Standard Oil Announcer" on one of San Francisco's streets. Standard Oil of California used the truck in the 1940s to promote its products throughout the state. *Standard Oil Company of California*

The Boiler Shop's polo team at the Richmond Refinery in 1921. Instead of riding horses, the men were mounted on bicycles, and they used the large mallets to push a puck, rather than a ball, around the court. *Standard Oil Company of California*

An interior view of the Crude Still Receiving House at the Richmond Refinery. The men on the right are taking samples of crude for testing as they reach the refinery via various pipelines from disparate fields, while the man on the left logs the results of their testing. *Standard Oil Company of California*

Refueling the *Southern Cross* as the aircraft is prepared at an airfield in Oakland for its historic trans-Pacific flight from California to Australia in 1928. Grateful for the opportunity to promote its aviation fuel, Union Oil Company supplied the gasoline used on the trip. *Union Oil Company of California*

Although there was no market for much of California's early natural gas production, some was sold to unexpected clients. For example, when Germany's *Graf Zeppelin* visited the West Coast in the late 1920s, the federal government prohibited the sale of helium to refill the gigantic gas bags. As a substitute, the Southern California Natural Gas Company supplied a mixture of 55-percent natural gas and 45-percent pyrofax, a commercial name for compressed propane gas. This photograph shows the *Graf Zeppelin* moored at the company's natural gas lease near Los Angeles. *PennWell Publishing Company*

McKittrick–Kern River

THE San Joaquin Valley, located in central California, is another of the state's major petroleum-producing areas. At the southern end of the valley is Kern County, an area that boasts several of America's greatest oil fields—McKittrick, Kern River, Midway-Sunset, Elk Hills—and a multitude of smaller pools, with tremendous oil and gas reserves. The McKittrick and Kern River fields were both uncovered in the late nineteenth century, while the Midway-Sunset Field and Elk Hills had to await the turn of the new century. The center of Kern County's prolific oil industry is its county seat, Bakersfield, a major regional oil-distribution point.

One of the giant producing areas of Kern County is the McKittrick Field, located in the foothills of the Temblor Range along the southwest border of the San Joaquin Valley, a few miles west of the town of McKittrick. For years prior to the California oil boom, when water was more valuable than crude, area residents joked that "when anyone tried to bore a well for water . . . they always struck oil." As early as the 1870s, residents of the San Joaquin Valley exploited McKittrick's asphalt beds for material to pave Bakersfield's streets.

Because most Californians were well acquainted with conventional mining methods, they knew how to recover the asphalt by sinking shafts into the exposed outcrops. However, one major difference between asphalt and gold mining was that in the former "the miners worked stark naked" because they quickly became covered with asphalt. While the men were underground, this presented no problem. But as a longtime resident recalled, they were so sticky that anything they touched became covered with asphalt. As a result, when the men came to the surface for their lunch break "'au natural,' clothed only in asphaltum that shown [*sic*] like lacquer," and sat on newspaper-covered benches in the company mess, the "newspapers . . . stuck to them like sticky fly paper." This created quite a scene when the miners returned to work with the paper "waving in the wind." "At the end of [a] tour," the observer said, "they were scraped with a case knife, or the wooden scrapers used on race horses, and washed in distillate" to remove the asphalt.

The asphalt was so plentiful that the Southern Pacific Railroad built a spur line to the town of McKittrick, originally called Asphalto, which became a major asphalt supply center. During the boom years, in spite of its relative isolation and shortage of drinking water, which had to be hauled in on the railroad for a dollar a barrel, the community boasted a refinery. The superintendent's house and office doubled as the U.S. Post Office, and there were a bunkhouse for the workers, a cookhouse, and a saloon.

The first development of the McKittrick Pool's petroleum reserves was started by Buena Vista Petroleum, which after acquiring a lease in the area in 1872, completed a few shallow wells to provide crude to mix with the brittle asphalt and thereby create a more usable product. The asphalt-crude mixture was placed in large vats for melting. When liquified, it was poured into boxes and sold as street paving.

Extensive development of the McKittrick Pool began after 1900 and continued rapidly for fifteen years. Eventually, approximately five hundred wells were drilled on 1,545 productive acres that stretched over an area four miles long and one-half mile wide. Most of the wells averaged between 800 and 1,400 feet in depth, and the producing sands varied from 100 to 500 feet thick. Strangely, given the shortage of surface water, a major problem encountered at McKittrick was the incursion of water. To combat this, compressors were installed and a million barrels of water were blown from the wells monthly for several years. Eventually the water table in the area declined, the compressors were removed, and heavy pumps were used on the wells.

Other pools of the McKittrick Oil Field are the McKittrick Front and the Cymric pools. The McKittrick Front Area, located between the McKittrick Valley on the southwest and the San Joaquin Valley on the northeast, was developed primarily by the Standard Oil Company of California. The discovery well was completed in July, 1916, as a ten-barrel-per-day producer. Most of the wells in this area did not have a large output, but those in the southern part of the pool proved to be extremely long-lived.

The Cymric Pool, also uncovered in 1916, was located just northwest of the McKittrick Front Hills. The first well, drilled by H. S. Williams, had an initial production of 150 barrels daily at 1,375 feet. By 1943, thirty-three wells had been completed, with initial productions like that of the discovery well. Strong gas flows were encountered, and one well blew out, with 70 million to 100 million cubic feet of gas flowing daily through the casing.

While the McKittrick Field was being developed in extreme western Kern County, another prolific find was made just to the northwest of Bakersfield along the Kern River. The initial discovery there, in what became the Kern River Field, was made by James and Jonathan Elwood in the spring of 1899, when they completed a well they had dug with a pick and shovel. When they reached a depth of forty-three feet, a strong seep of natural gas forced the Elwood brothers to install bellows for ventilation. Even so, when they attempted to deepen the hole, so much oil and natural gas seeped into the well that, at about the sixty-foot level, work became impossible. Later the Elwoods employed a steam-powered drilling rig and hired Milton McWhorter as the driller to complete the well. After deepening the hole by approximately forty feet, the drill bit found crude. The well's initial production was as much as fifteen barrels daily, which created a major storage problem. With nothing else available in which to store the crude, the Elwoods and McWhorter gathered all the "whiskey barrels, . . . milk cans, kerosene cans, beer kegs" they could find—and anything else that would hold the oil—and filled them.

At first people who lived in nearby Bakersfield did not take the discovery seriously. However, as word spread, oil men from throughout the region flocked to the strike. Thousands of acres of "worthless . . . sandy, parched and stony" land could be bought for between $2.00 and $2.50 an acre prior to the strike. However, after the discovery of crude, the price jumped to $5,000.00 an acre, with some tracts bringing as much as $11,000.00 an acre. Hundreds paid $10.60 for a round-trip ticket on a special sightseeing excursion train that the Southern Pacific routed from San Francisco to the Kern River Field.

The success of the Kern River discovery was assured when the San Francisco Central Power and Light Company, which provided natural gas and electricity to many San Francisco businesses, lost its supply of the anthracite coal used to power its generators. At first the company attempted to replace the hard coal with soft coal, but the resulting air pollution was so great that the San Francisco Board of Supervisors ordered that another source of energy be found or the generators be shut down. Because the field was near two railroad lines, which made shipment to market easy, the newly discovered Kern River oil deposits were a logical choice.

However, first a major problem had to be overcome. The Kern River oil was "like molasses"—so thick that it could not easily be unloaded from the railroad tank cars used to transport it to the generating plant. After some experimentation, J. W. Pauson succeeded in designing what became known as a "goosebill," which solved the problem. Pauson's invention was a length of one-inch pipe that was flattened at one end. Steam was injected into the pipe, the pipe was placed in the heavy oil, and the heat caused the heavy Kern River oil to flow freely. Within a short time the San Francisco Central Power and Light Company's generators were being fueled by Kern River oil. Just to the south of the pool and the river, and straddling the Southern Pacific Railroad, was Kern City. A progressive community, it counted a population of approximately two thousand by July, 1901. Because of its railroad outlet, Kern City was the obvious shipping point for Kern River crude.

Within a short time, oil men had completed several additional wells in what became the Kern River Area of the Kern River Field. Most were primitive, with the oil being scooped from hand-dug sumps. Pauson's use of steam to thin the heavy, low-gravity Kern River crude eventually was applied to many producing wells in the field. In 1901, J. W. Goff ran a steam line into one of the local wells. An air line then was run through the steam line. The steam heated the air, the hot air heated the crude, and the oil flowed more easily. Since Goff's invention came during a time of declining oil prices, though, most oil men did not deem it practical to go to the extra expense of adding it. Later, steam injection greatly increased the output of the Kern Field.

In the Kern River Area most of the output came from small wells. No gushers were ever located, and only a few of the wells flowed at more than 500 or 600 barrels of crude daily for a very long period. Between the time of its discovery and July 1, 1938, the Kern River Area produced more than 300 million barrels. On September 1 of that year, daily production amounted to 9,130 barrels of oil from 2,049 wells spread over 9,693 proven acres.

Like that at McKittrick, the discovery at Kern River was plagued by water incursion, especially in the early 1940s. The Getty Oil Company, which controlled thirty-five hundred acres of production in the pool, installed turbine water pumps that handled 10,000 barrels per day to solve the problem. By this method, Getty eventually was able to restore production in a thousand wells.

It was not unusual for rival factions to defend violently what they considered their drilling sites. Once Frank Hill took a Union Oil Company drilling crew to a lease near Maricopa in Kern County. When they arrived at the lease, "they found themselves looking into the barrels of shotguns." The original occupants "told us to get the hell out of there, and get out fast," Hill later recalled. "Well, we were drillers and not gun fighters, so we got the hell out, like they said." "Later, we went back and drilled," Hill continued, "but that was after our land boys had leased the acreage all over again from the crowd that ran us off in the first place."

As more and more oil men rushed to the Kern River Field, Bakersfield, the nearest community of any consequence, quickly became a boom town. "Saloons, dance halls, and sporting houses ran wide open," as "men frantically leased and re-leased land." Claims were "staked and jumped by the dozens," while oil men fought "with fists and sometimes with their guns to maintain their titles."

Well on its way to becoming a major petroleum center, Bakersfield boasted a population of 12,727 by 1910. The town received a tremendous boost when, between 1902 and 1909, two major pipelines were constructed, one by the Standard Oil Company and the other by the Associated Oil Company and the Southern Pacific Railroad, to link much of the Kern County–San Joaquin Valley oil fields to West Coast markets. At the time, it was a tremendous undertaking at an enormous cost. Standard Oil spent three million dollars on its line alone. However, the increased marketability of Kern County's crude so stimulated production that by 1909 California was the nation's leading oil-producing state.

Indeed, the flow of crude outstripped demand, and the price of California oil dropped. In an effort to increase prices, 150 companies operating in the San Joaquin Valley formed the Independent Oil Producers Agency (IOPA). When neither the Standard nor the Associated oil companies would pay the price the IOPA set, its member companies negotiated a decade-long contract with the Union Oil Company, which agreed to buy up their production at the same price it received for its own crude. To fulfill its part of the bargain, the Union Oil Company built another $4.5 million pipeline from the valley area to Port Harford on Morro Bay.

Completion of the Union pipeline gave Kern County oil men several options in marketing their production. After the line's opening, oil could be delivered to the Pacific port at a cost of 12.5¢ per barrel. Once the crude reached the coast oil men could store their oil at a cost of 2¢ per barrel annually in open pits, or 1¢ per month in covered steel or wooden tanks. Or they could transship the oil to other California ports at prices that varied between 10¢ per barrel to San Diego or San Francisco and 20¢ a barrel to Eureka.

Although it was a thriving oil community, Bakersfield tried to avoid some of the pitfalls of other oil boom towns. So that culture would not be ignored, a Bakersfield Opera House was opened, and the Ziegfeld Follies was brought to the community. On another occasion, a 150-mile automobile race was promoted through the surrounding oil fields. Of the three events, the automobile race, held on the Fourth of July, 1911, proved the most popular. The promoters not only persuaded the Southern Pacific Railroad to halt rail traffic so that the course could cross its tracks, but also convinced local oil companies to employ more than two hundred men to make necessary improvements to existing roadways and to build roads where there were none along the race circuit. In addition, to keep the dust down, nearly every oil company water wagon in the county was used to moisten the dirt track. The event attracted several nationally prominent racers, including at least one car that claimed to be capable of reaching a speed of one hundred miles per hour. On July 4, 1911, ten thousand people lined the race circuit, and at 11:05 A.M. the first racer was flagged off. Nearly three hours later, the competition was over. Harvey Herrick, driving a National automobile won; Bert Dingley, second, drove a Pope-Hartford; Frank Seifert finished third in a Mercer.

While much of the early development was in the Kern River Area, the Kern Front Area of the Kern River Field was opened in 1915, about ten miles north of Bakersfield, on the east side of the San Joaquin Valley. Stretching along the western flank of a broad structural arch (the Kern River arch), the pool was uncovered by the Standard Oil Company of California. Its discovery well, the Fee No. 1, was completed at a depth of twenty-five hundred feet with an initial flow of 500 barrels daily. By January 1, 1941, the productive acreage of the Kern Front Area had been expanded to over three thousand acres, about 80 percent of which had been developed. Cumulative production was pegged at 45.5 million barrels of crude by the beginning of 1941. In fact, recovery as of January 1, 1941, stood at approximately fifteen thousand barrels of oil per acre.

Several other nearby pools were uncovered along the arch. Northwest of Bakersfield and south of the Kern Front Area, the Fruitvale Pool was uncovered, and northwest of the Kern Front Area the Poso Creek, or Premier, Pool was discovered. The Poso Creek Pool and nearby Mount Poso were remote, even by Kern County standards. Much of the productive acreage of the field was controlled by J. Paul Getty, and as late as 1926 only "four tin shacks" marked the field, which at that time "was seldom visited." If shunned by humans, Poso Creek was home for thousands of rattlesnakes. One of the shacks there had a wall covered with more than two hundred rattles taken from snakes killed nearby. In 1928, the Poso Creek Field still was producing forty-five barrels of crude per month.

Although the McKittrick and Kern River fields were prolific producers, they were caught in a circle of increasing production, decreasing demand, and lower prices. As the price of crude began to drop in the 1920s, many California oil men faced a dilemma. To offset the drop in price, the Los Angeles oil operators, who were affiliated with the Chamber of Mines and Oil, decided on July 21, 1921, to reduce wages one dollar per day. The decision was announced to the Kern County oil workers on August 2. Taking effect on September 1, the cut in salary pegged the going daily wage for a roustabout at $5.00; a teamster or light truck driver would be paid $5.25; pumpers, stablemen, light tractor drivers, and machinist helpers would earn $5.50; boiler washers, second engineers, and six-horse teamsters would be paid $5.75; heavy truck drivers, first engineers, and skilled mechanics would receive $6.00; drillers were offered $9.00; and other employees would be paid between $6.50 and $8.25 daily.

Irate over the wage decrease, the International Association of Oil Field, Gas Well, and Refinery Workers, A.F.L., went on strike on September 11, 1921. Centered in the San Joaquin Valley area of Kern County and extending northward into the Coalinga Field, the walkout stopped most drilling activity in the region. In all, nearly eight thousand men left their jobs. When the operators attempted to bring in strikebreakers three days later to resume production, two thousand strikers forced the Southern Pacific train hauling the non-union workers to halt at Pentland Junction near Maricopa and made it return to San Francisco. Although the confrontation had been heated, it had been bloodless. Nevertheless, fearing another face-off between union and non-union workers, the sheriff of Kern County banned the sale of guns and ammunition.

Many residents and businessmen in Kern County backed the strike. Rallies were held to offer support, and many oil-town businesses extended credit to the union workers. To prevent violence and ensure continued local cooperation, the International Association of Oil Field, Gas Well, and Refinery Workers called for volunteers with prior military experience and organized them into "patrols" to "maintain law and order." At the same time a Law and Order Committee was formed to police the fields. One side effect of the crackdown was that it forced most of the region's gambling dens and saloons to close and most bootleggers to flee.

To combat the strike, the operators formed the Oil Producers Association of California (OPAC). Taking a hard line, the producers decided to break the union by keeping the men out of work and going to court to evict striking workers from lease houses. To offset the economic hardships faced by the strikers, the union began paying strike funds. Single men received ten dollars; married men, fifteen dollars. Much of the money came from local merchants.

The seeming inability of the two sides to compromise hardened feelings. It appeared that open fighting might erupt when several armed guards of the OPAC challenged some of the strikers' guardposts, which had been established to keep non-union workers away from the wells. On October 13, the strikers appealed to the governor of California to send in the National Guard to prevent bloodshed.

Finally, though, the potentially dangerous situation was defused. As early as November 1, the strikers began voting on whether they should end the walkout. The initial attempts to end the strike were rebuffed by the OPAC, however, which declared that the proposals by "the various locals that the strike be called off will have no appreciable effect upon the plans of the Oil Producers Association and member companies." Two days later, on November 3, 1921, the union men accepted the dollar cut in pay and voted to return. Concurrently, the OPAC agreed not to lengthen working hours. Within a short time the fields were back to normal.

Violence had been avoided, but within a year Kern County was jarred by the sudden appearance of the Ku Klux Klan. In early February, 1922, the KKK issued a warning to local "gamblers, gunmen, bootleggers, loafers, lawbreakers of every class and description" to obey the law or "Beware!" Afterward, in a parade held in Taft, the klansmen drove through the streets with guns pointing from automobile windows. There were reports from both men and women of beatings administered by klansmen. Several area residents fled, and others were tarred and feathered, before local authorities made a local membership list of the Klan public. Stripped of its protective layer of secrecy, the KKK saw its power fade.

After the 1920s the development of Kern County's oil reserves continued. In 1943, for example, 126 oil wells were completed in the Kern Front Field. Much of the later drilling was deep, with wells in excess of 12,000 feet. By 1946, Kern County had more deep wells, 96, than any other California county. In fact, Kern County had more than any other county in the country. Pacific Western's National Royalties No. 1, a Kern County wildcat, became the world's deepest well on December 31, 1945, when it reached 16,668 feet.

This continued development enabled Kern County to maintain its leadership role in both the American and the California petroleum industry. Eventually, the Kern River and the McKittrick-Cymric pools became two of the nation's greatest oil fields, ranking sixteenth and sixty-fifth respectively among America's most prolific oil fields at the end of 1949. By the end of the first half of the twentieth century, the two fields had a total combined output of 422,375,000 barrels of crude, and these were not the only major fields in Kern County. Indeed, the massive Midway-Sunset and the controversial Elk Hills fields both brought national attention to the county.

By 1887 shallow wells were replacing the asphalt mines in the McKittrick area. The pool in the foreground contains oil, not water, but during some periods water was more valuable than oil to area residents. *American Petroleum Institute*

A panorama of the McKittrick Field in 1900. The desolate location of the discovery retarded the pool's development. *Standard Oil Company of California*

Drilling at the bottom of arroyos in the McKittrick area could be dangerous, for although the region suffered from a lack of surface water, sudden rainstorms could send torrents of raging water sweeping through low-lying areas. *Standard Oil Company of California*

The McKittrick Field, 1900. Apparently the well has been completed, since the walking beam has been locked in the raised position and casing has been staked alongside the derrick in preparation for pouring cement. *Standard Oil Company of California*

Because high winds often swept through the McKittrick Field, a wooden wall has been built on one side of this steam boiler for protection. *Standard Oil Company of California*

A pipeline carrying water to a steam boiler at McKittrick in 1908. Ironically, although above-ground water was scarce at McKittrick, the field was plagued by water incursions below ground. *Standard Oil Company of California*

A cable tool rig belonging to Standard Oil Company of California at McKittrick in 1900. The boiler is to the right of the derrick. The small shed to the left of the derrick covers the band wheel. The two boxes running between the shed and the derrick house the calf rope, on top, and the bull rope, on the bottom. *Standard Oil Company of California*

The Southern Pacific Railroad's spur line into McKittrick allowed supplies to be shipped into the field inexpensively. Because of the dry climate that retarded spoilage, the material usually was simply stacked by the side of the tracks until it was needed. Notice the McKittrick Oil Exchange in the background, which offered area oil men the latest quotes on oil prices from the telegraph stretching alongside the railroad tracks. *Standard Oil Company of California*

The town of McKittrick in 1900. Its relative isolation in the foothills of the Temblor Range made it difficult to attract workers to the wells, when they could easily find work in more pleasant surroundings in the booming Los Angeles Basin. *Standard Oil Company of California*

A typical boom-town bunkhouse in the McKittrick Field. Because of the hot, dry climate, the roof has been covered with overlapping boards instead of sheet-iron, which would have made the inside temperature unbearable. While the bottoms of the windows are open to catch the breeze, the top halves are shaded to keep out the sun. *Standard Oil Company of California*

A group of McKittrick oil-field workers and their families on a picnic at nearby Sulphur Springs. Although the water had a sulphur smell, it was one of the few attractions in the rough terrain of Kern County. Note the large, almost wagon-type wheels on the automobile, which were necessary to negotiate the rough countryside. *Getty Oil Company*

An aerial view of McKittrick in 1937. The Southern Pacific's rail lines are running almost due northwest before veering sharply toward the northeast. The town lies to the southwest of the tracks, and the wells stand on the high ground northwest of the community. *Standard Oil Company of California*

Left: Hauling barrels of water at McKittrick. Water had to be carried to the well sites for the workers. Note the dipper hanging by the side of the driver. *Standard Oil Company of California. Right:* Three different types of derrick rigs standing beside one another in the McKittrick Field. The one on the right is a wooden derrick, the one on the left is a permanent steel derrick, and the one in the background is a portable steel derrick. *Union Oil Company of California*

A portable drilling rig operating in the McKittrick Field in 1947. In the post–World War II era there was a revival of activity in the region. The crew is preparing to attach the next length of pipe, to the right of the derrick, to the one currently in the hole. *Standard Oil Company of California*

The Kern River Field in 1900. Although the pool had been located only the year before, by this time it was already well developed. Notice the rivulet of crude in the right foreground. *American Petroleum Institute*

A Kern Oil Company well in the Kern River Field. Notice the pipe running out of the bottom of the storage tank to the small ledge on the right of the photograph. Tank wagons, such as the one shown in the picture, would stop under the pipe to fill before carrying the crude to Bakersfield for refining. *Getty Oil Company*

The double railroad track spur through the Kern River Field. Because there was not enough storage space, pipe was simply unloaded from flatcars and stacked beside the track until it was needed. *Getty Oil Company*

In 1928 the Kern River Field was struck by a freak snow and windstorm that damaged many of the pool's derricks. These men are repairing area roads after the storm. Note the wooden culverts used to carry off excess water. *Getty Oil Company*

Rees Store and Post Office at Oil Center in the Kern River Field. Typical of boom-town constructions, the post office is a clapboard, false-front structure, with a wooden sidewalk. One man in the center is proudly showing off his young child. *Getty Oil Company*

Getty Oil Company's Oil Center office in the Kern River Field in 1925. Surrounded by a spacious roof that shaded the windows, the building had an open foundation to allow circulation. Both features were necessary in the hot climate of Kern County. *Getty Oil Company*

Ten-mule teams hauling tank wagons of Kern River crude to the Santa Fe Railroad at Bakersfield in 1899. This was one of the first loads of Kern River oil delivered to the railroad for use in its locomotives. Each team is pulling two tank wagons. *Pacific Oil World*

A group of Kern River workers in front of company bunkhouses. Each two-story structure had a large, wrap-around porch for each level. Along with the large overhang of the roof, the porches were designed to provide shading from the sun and to allow a place for the workers to relax in the cool of the evening. *Atlantic Richfield Company*

A panorama of the Kern River Field taken from the bluffs overlooking the Kern River in 1923. Note the two parallel lines of derricks in the center of the photograph with the power sheds of both lines on the inside of the square. The owner of the lease had apparently drilled a series of wells on the very edge of the property to prevent any nearby well from pumping oil from beneath his lease. *Pacific Oil World*

While there are several steel storage tanks visible in this photograph of the Kern River Field, oil from the well in the right foreground is flowing into an open pit. *Atlantic Richfield Company*

By the 1920s the automobile had replaced the horse and mule power used at first in the Kern River Field. Note that there are eight vehicles and twenty-four men visible. With three drivers for each automobile, the vehicles could be kept in operation twenty-four hours a day, with each driver working an eight-hour shift. *Atlantic Richfield Company*

Crude from the Kern River area was so thick that heaters, such as the one shown here, were used to thin the oil so that it could be transported through pipelines to markets. *Atlantic Richfield Company*

The Pan American Petroleum Company's gathering system in the Kern River Field. Pipelines from Pan American leases in the region all converged on the plant, where the crude was cleaned and heated so that it would flow more easily. *Atlantic Richfield Company*

Left: A Lufkin pumping unit in operation on the Richfield–Pan American petroleum companies' Kern No. 29 in the Kern River Field. The small electric motor on the left of the pump provided the necessary power to operate the system. *Atlantic Richfield Company. Right:* A gun perforator being lowered into a well near Bakersfield in the Kern River Field. The perforator is lowered into a well that has had casing already set in order to open to production an area that is above the bottom of the hole. When the perforator reached the desired depth, it was exploded, and steel bullets were shot through the casing so that oil could enter the pipe. *Atlantic Richfield Company*

Railroad loading racks and oil storage tanks at Bakersfield, about 1918. *Atlantic Richfield Company*

The Lakewood Oil Company supply yard at the Bakersfield railroad depot. Much of the equipment was loaded directly off the flatcars onto wagons, for transport to drilling sites. The wagons had no seats for drivers; instead, the driver rode the left rear horse of the hitch, while the dog in this photograph appears ready to ride on top of the load. *Security Pacific National Bank Collection, Los Angeles Public Library*

The corner of Fifth and Fresno streets in Bakersfield. During the oil boom the community's population mushroomed. To handle the influx of oil men, the town expanded rapidly—so rapidly that the provision of many amenities, such as paved streets, could not keep pace. *Atlantic Richfield Company*

Because of its proximity to the Kern River Field, Bakersfield became a major oil center. Shown here, in a 1916 photograph, is the Bakersfield Refinery, which handled much of the crude produced by nearby wells. Prior to the completion of the Associated Oil and Standard Oil companies' pipelines, much of the crude was marketed locally by salesmen in wagons like the one in the foreground. *Atlantic Richfield Company*

A restored cable tool rig on the grounds of the Kern County Museum in Bakersfield. Sponsored by the Petroleum Production Pioneers, the rig has become the gathering place for the Pioneers during their annual barbecue. The walking beam, clearly visible protruding from the well house, was connected to the band wheel by the stirrup, a large circular piece of metal. The steam engine rotated the band wheel, which raised and lowered the walking beam, which in turn raised and lowered the drilling cable and the bit in the hole. As the bit was raised and lowered in the hole, it pounded its way through the earth. *PennWell Publishing Company*

An interior view of the Richfield Oil Company's laboratory at Bakersfield. Samples of oil from many nearby wells were brought here for testing to determine their gravity and paraffin content. *Atlantic Richfield Company*

Left: By the late 1920s, the Hansen electric motor had replaced the early method of welding with acetylene in the San Joaquin Valley. The electric current, supplied by the generator mounted on the wagon, was applied to the pipe to form the welds. *PennWell Publishing Company. Right:* Pipeline crew bending a section of pipe on the San Joaquin Valley pipeline. The men would line up on the pipe and bounce up and down, slowly bending the pipe to the correct angle. *PennWell Publishing Company*

An oil-field bridge in the San Joaquin Valley that collapsed as a truck loaded with pipe attempted to cross, 1909 or 1910. Workers have stabilized the truck with cables to keep it from falling off the debris. Before the vehicle can be removed, however, it will be necessary to move the pile of pipe and brace what remains of the bridge. Notice that the truck is chain driven and that the solid tires are notched to give them better traction. *PennWell Publishing Company*

A team of horses hauling boilers across the desert during the construction of the Producers' Oil Company's pipeline from Kern River to the Pacific coast. The boilers were used in the pumping stations along the line to keep the crude flowing. *Union Oil Company of California*

Two of the tractors used to connect the oil fields of the San Joaquin Valley with outlets on the west coast. The huge radiators on the tractors were made necessary by their operation in the hot, dry climate. The driver sat near the rear of the tractor, above the Caterpillar treads, and steered the machine by a steering wheel connected to the single wheel in front. Both tractors were equipped with canvas coverings to protect them from the weather. Each machine was capable of pulling several trailers, depending on the job. The wide wheels on the trailers were necessary to prevent them from sinking into the sand or soft ground when heavily loaded. *PennWell Publishing Company*

A 1913-model, five-ton, chain-driven White truck, used to haul pipe across the Mohave Desert to the west of Bakersfield. Although the vehicle was built to stand the heavy oil-field workload in the heat of summer in the desert, it did not offer the driver any protection from the elements. *White Motor Corporation*

Once the pipelines converged on the coast near San Luis Obispo, they were extended offshore, where they could empty crude into tankers for transportation to other markets. To lay the submarine pipeline, the Standard Oil Company first joined the pipe together on a system of rails and dollies. Once the pipeline was completed, it was pulled into the sea. *PennWell Publishing Company*

Automobile racers were sponsored by several California-based oil companies, and the sport was extremely popular among most oil-field workers. One race held on July 4, 1911, at Bakersfield offered a first-place prize of two thousand dollars, quite a bit of money for the day. Here a racing-car driver sponsored by the Richfield Oil Company poses, as the mechanic fills the engine with Richlube Motor Oil. *Atlantic Richfield Company*

An aerial view of the Kern Front section of the Kern River Field, taken in 1930. By this time the pool had been extensively developed. Notice the Kern River curving through the lower left of the photograph. *Atlantic Richfield Company*

A drilling crew eating lunch on a well site in the San Joaquin Valley in the early 1950s. As in many other producing areas in California, oil men continued to drill in the San Joaquin Valley for years after the initial strike. However, safety precautions were far stricter than in the early part of the century. Note that each man has a steel helmet for protection from objects falling from the derrick. In addition, the later-day roughnecks had many more comforts than their earlier counterparts, such as the lockers lining the wall of the dog house. *Standard Oil Company of California*

Midway-Sunset

IN 1901 another Kern County pool was uncovered to the south of McKittrick, about a mile southwest of present-day Taft, in what eventually became the huge Midway-Sunset Field. Although the Midway-Sunset discovery evolved into a conglomeration of separate areas, the initial strike was made in the Twenty-Five Hill region, and intensive development of the area started in 1907. Productive acreage was expanded for five miles along the northeast and southwest flank of the Spellacy anticline.

Initial production in the Twenty-Five Hill area ranged from ten to five hundred barrels daily from each well, with a few exceptional wells reporting production approaching ten thousand barrels a day. By 1917 approximately six hundred wells had been drilled along the Spellacy anticline; however, only four hundred proved to be producers. Their combined output approached four thousand to five thousand barrels daily.

Those Midway-Sunset wells which were completed for high crude production generally were accompanied by a tremendous natural gas flow. The results were some of California's most spectacular gushers. The most famous of all was the Lakeview No. 1. Spudded in about a mile and a half north of Maricopa on January 1, 1909, by the Lakeview Oil Company, the well gushed at an estimated sixty thousand barrels per day in its first three months.

Originally, the lease had been purchased by the Bakersfield city attorney in 1901 at a cost of five dollars an acre. However, after drilling a dry hole, the attorney sold the lease to Julius Fried. Although he was aware of the earlier duster, Fried recalled the secret to successful drilling in the area told to him by an old-time oil man. Drill where red grass is found, the pioneer had declared. In the dry climate, he had insisted, grass would grow only where moisture was retained in the soil, usually over an underground fault. Thus the presence of grass generally indicated a potential petroleum producing formation, and as the weather got hot, the grass turned red, which made it easy to see.

Believing the story, Fried spudded in a well on a patch of red grass on the Lakeview property. Just as the old oil man had predicted, the site was directly over a narrow bed of oil that, though only a few feet wide, was a mile long. Fried and his partners in the

Lakeview Oil Company, who had bought in at a cost of fifty cents a share, pushed their hole to the 1,655-foot level before financial problems forced them to seek additional money.

The Union Oil Company, which wanted to build storage tanks on the property, agreed to complete the well in return for 51 percent of the Lakeview Oil Company stock. Fried agreed, but work on the Lakeview No. 1 proceeded slowly because Union Oil manned the rig only when it could spare workers from other wells. Consequently, it was not until March 15, 1910, that the bit found oil.

Even then the crew did not realize that production had been found. Instead, Roy McMahon, the driller, thought that the bailer was stuck in the hole. When he informed production superintendent Walter Barnhart of the problem, Barnhart told McMahon to jerk the cable up and down in the hole to try to free the bailer. McMahon started jerking the cable and was startled when the bailer was blown over the well's crown block and a column of oil and gas shot out of the wellhead. The wooden derrick was quickly destroyed, and spewing sand covered the engine house, bunkhouse, and coal shed. The crude was thrown so high that it fell to earth miles from the well site as the "gigantic fountain sent a shower of oily spray, [and] spread a coat of black over gray sagebrush and brown, dry ground." Thousands of barrels of oil poured from the hole, eventually forming a small river, dubbed the "trout stream," which headed downhill toward Buena Vista Lake, eight miles away. The huge black stain caused by the spewing crude could be seen twenty miles away, "standing forth like a blot of ink upon the silky sheen of the slope."

Hurriedly, Frank Hill, Union Oil's director of production, rushed to the site to take command. To stem the stream of oil, he ordered earthen dikes constructed in its path. Such a project required more men than were available at the well site, so Union Oil hurried in workers from as far as three hundred miles away. "For weeks" the men toiled, "covered from head to foot with the black, sticky stuff." "Standing in a shower of hot oil that caused the skin to blister and peel off wherever it stuck," they tried "to smother it with caps and rafts weighing many tons, attempting to spread aprons over the fountain, to tie it down with masses of steel."

Twenty large sumps were dug to catch the spewing crude. In addition, four hundred men built another sandbag and sagebrush dam around the entire well to hold runaway oil, which at its peak was flowing at an estimated 90,000 barrels daily. Within two and a half months, 3.5 million barrels had escaped. Although the flow of crude had been contained, it could not be stopped. In an effort to control the well, the workers built a "high and wide stockade of alternate rows of brush and sand" around the mouth of the well. "Heavy planks on the outside completed the corral." As a roof, "an immense raft of timbers sixteen inches square" was constructed. Then when the flow of crude slackened a little, the roof was thrown "over the mouth of the well . . . and anchored . . . with immense chains on the four corners."

When the pressure built up inside the well, the spewing crude pushed the roof "high up, to the limit of the chains" before the "sixteen-inch timbers were scattered to the four winds." Fortunately, the roof held long enough for the stockade to fill with crude and choke the flow of oil. On September 9, 1911, the Lakeview No. 1 caved in on the bottom and sealed itself. During its uncontrolled spewing, the well had produced an estimated nine million barrels of crude. Of that output only four million barrels were saved.

Another wild well was brought in by the Pacific Crude Oil Company in March, 1912, about seven miles from Taft, with an estimated flow of ten thousand barrels daily. Everything was normal for about eight hours, then the tremendous gas pressure forced its way past the wellhead and threw a geyser of oil three hundred feet into the air. Shortly afterward a fire broke out in the engine house of an adjoining lease. The flames quickly ignited the oil-soaked ground. Suddenly, as the fire burned toward the Pacific Crude Oil Company well, "a little line of light trickled up to the oil geyser." "There was a roar, a deadly black volcano of smoke and flame shooting heavenward."

As "men ran out of the danger belt to escape the shower of fire drops," the fire continued to spread. A sump hole, about seven hundred yards from the well, that contained five thousand barrels of crude burst into flames. Soon neighboring leases and equipment were caught in the inferno. For a while it was feared that the nearby "gigantic oil tanks" would catch fire, but several hundred men, working behind asbestos shields, turned the flames back. As one witness reported, "The sun was shut from the field by the clouds of dense black smoke billowing in every direction." "Now and then" he continued, "smoke rings, as though puffed from the mouth of a giant, would shoot upwards, and would travel perhaps for ten minutes high in the heavens before they would gradually shift into undefined forms." When night came, the flames were so bright that "people for two or three miles around could sit on their front porches and read quite easily in the glare."

All nearby boilers were utilized in an effort to extinguish the fire. Several times the well sanded up, but before the workers could secure it the gas pressure would build up and throw the stream of crude and gas into the air once again. Finally, after six days of almost constant work, which so exhausted some of the men that they had to be hospitalized, the steam from the boilers began to force "the lurching fire pillar down, lower and lower, until it was finally quenched." On the morning of the seventh day the well was capped, but not before the fire had consumed between sixty-five thousand and one hundred thousand barrels of oil.

In 1929 the Union Oil Company uncovered a deeper paying horizon along the Spellacy anticline, when the Williams No. 1 was completed at 1,494 feet for 250 barrels daily. The Williams Area, as the region became known, was slowly developed until 1936, when the Chanslor-Canfield Midway Oil Company brought in a 770-barrel-per-day well. Afterward oil men flocked to the scene, and many wells were drilled on small leases. Because of the concentration of derricks, the region was nicknamed Little Signal Hill.

Later the initial Midway-Sunset discovery was expanded to the northwest by the development of the North Midway Area. This proved to be one of the most productive regions of the entire field, and 10,000-barrel-per-day wells were not uncommon. Through a geologic quirk, however, the wells along the southwest and northwest edges of the pool produced an average of only 5 barrels daily.

Another prolific producing area of the Midway-Sunset Pool was the Republic Area, which covered approximately ten square miles just to the south of and adjacent to the town of Fellows. Most of the early production in this region came from the Buttress Sand, a shallow sand zone that produced many spectacular gushers from around 800 or 900 feet. Peak output of the Buttress Sand occurred between 1911 and 1914 and between 1920 and 1921.

The Midway-Sunset Field saw the most active use of rotary rigs up to that time in California. As early as 1908, Standard Oil Company moved several rotary rigs and their crews from Louisiana to the Midway-Sunset Field. The initial effort proved so successful that in 1911 the Union Tool Company of Los Angeles started constructing rotary rigs to meet the new demand. Standard Oil Company purchased the first two and immediately shipped them to Midway-Sunset. Not all of Midway-Sunset's wells were rotary drilled, though. Many companies still relied on cable tool rigs. Cable tool rigs were not as mechanized as the rotary rigs and depended more on the ability of the driller and crew than sophisticated equipment.

For example, on one well Standard Oil drilled in 1910, immediately west of Taft, the tools stuck in the hole at a depth of 546 or 547 feet. All conventional efforts to free them failed, but a string of sixteen-inch casing had been set to the 545-foot level, and John Stuck volunteered to be lowered into the hole to free the tools. To make sure there was enough oxygen in the well to keep Stuck from suffocating, workers first dropped a torch into the hole. When the torch continued to burn, Stuck "removed his jacket and tied a line under his arms." Lowered feet first, he made it safely to the tools, but there was not enough room for Stuck to bend over and reach them with his hands. Undaunted, he kicked the bit with his feet. Stuck was pulled quickly to the surface, closely followed by the bit.

Stuck's action was not unique. Several innovative methods were developed to remove stuck tools in the California oil fields. Once at Santa Fe Springs, a string of tools had been stuck in a well belonging to the George F. Getty company for several weeks. The crew had about given up hope of clearing the hole, and because Santa Fe Springs was a town-lot field with the wells close together, neighboring wells were draining off oil from beneath the Getty lease. Unless something was done quickly, the Getty property would be worthless. When confronted with the problem, J. Paul Getty, as he later recalled, "went to the cemetery and bought a six-foot round granite . . . [shaft], then had one end cut to taper." "I took this back . . . and asked the driller to throw it down the hole," Getty continued. The granite pushed the stuck tools out of the way, and the method became known as "the Paul Getty Special."

As of June 1, 1941, daily allotment for the Republic Area was set at 528 barrels, and cumulative production totaled approximately 11 million barrels, or an average of 137,000 barrels of crude per acre. The Republic Area was only one of two giant pools in the Midway-Sunset Field, though. In 1909, the Buena Vista Hills Area was located by the Honolulu Consolidated Oil Company. Actually a separate pool, the low-lying Buena Vista Hills Area paralleled the original discovery at Twenty-Five Hill to the northeast. Although a portion of the pool was kept from production and formed the Petroleum Naval Reserve No. 2, approximately 50 percent of the productive acreage was open to private oil men.

The discovery well, the Honolulu Consolidated Oil Company No. 1, blew in out of control as a gasser from a depth of 1,608 feet, but, when deepened to 2,540 feet, flowed at 2,500 barrels of oil daily. The tremendous natural gas pressure was to be a problem in many Buena Vista wells. In one instance, an Eagle Creek Oil Company well near Fellows blew in with such force that it spewed millions of fossil seashells out of the well-head. The ground around the well turned white as it was covered with the shells.

Although the oil men knew of the pressure, and took all available precautions to prevent blowouts or fires, the danger was never eliminated. When the Kern Trading & Oil Company's (K.T.&O.) No. 21 blew in wild, it threw an estimated twenty-five thousand barrels of crude into the air daily. A careless pumper ignited the runaway well with a cigarette, and as one witness recalled, the "belching flames . . . [made] it look as if the fires of Hades" were loose. Ten thousand dollars' worth of crude was lost daily to the flames, which could be seen twenty miles away.

An attempt was made to douse the fire with steam and mud. Shields to protect the crews were hauled into place, and a huge flow of steam from eight-inch pipes was played on the well, followed by a deluge of mud. The fire was extinguished briefly, but the well was reignited by the still-burning sump. Another attempt to stop the fire was made using steam, mud, and dynamite, but it was not until November 3, 1913, sixteen days after it first caught fire, that the well was finally snuffed out by ten thousand barrels of the chemical used in fire extinguishers. When the flames were at last extinguished, a crater thirty feet wide and forty feet deep was all that remained.

Development of the Buena Vista Hills Area was sporadic. Although it started immediately after completion of the discovery well, there were several instances when drilling activity practically ceased, particularly between 1919 and 1921 and after 1929. Yet, by January 1, 1942, 1,367 wells had been drilled in the field, with 398 of them abandoned as unprofitable. Average production of the Buena Vista Hills Area stood at 15,154 barrels daily in December, 1940, with a cumulative production of 297,116,866 barrels as of that date.

To handle the output of the region, the Santa Fe and Southern Pacific railroads in 1901 jointly constructed the Sunset Railroad stretching southward from Shale through the field to a junction with the larger rail lines. Eventually, a network of small settlements was established along the west side of the forty-six miles of track. One of the most important of these communities was Taft, located on the western edge of the pool about halfway between the northern and southern ends. Originally founded in 1908 when the Southern Pacific and Santa Fe railroads completed a spur line about seven miles northwest of Maricopa, it was initially named Siding Two. Another nearby collection of shacks was started and named Moron. Eventually, in 1910, Siding Two and Moron were combined and incorporated as Taft, with a population of 750.

There was no local supply of drinking water. All potable water was hauled in at a cost of two dollars a barrel. Glasses of water sold for five cents. Water was so scarce that some oil companies provided their employees condensed water drawn from the coolers on the steam boilers to drink. Obviously it was too expensive to wash in "good" water, but several nearby sulphur water wells supplied sufficient "bad" water in which to bathe. Because of the water shortage, Midway-Sunset workers cleaned their clothes by soaking them in distillate and then placing them in blow-off boxes at the end of steam lines. Steam from the boilers worked with the distillate to remove the dirt and grime.

In spite of the hardships, the intense oil development brought thousands of workers rushing into the area. When they arrived in the field, they found that room and board cost seventy-five cents a day. Wages paid the workers varied. A man with a horse and wagon usually commanded five dollars a day. Drillers also earned five dollars a day, and tool dressers, whose task it was to look after the drilling tools, earned four dollars. Contractors charged one dollar a foot for most early-day drilling; however, if the hole encountered sand heavily saturated with water, they charged an additional ten cents per foot.

Most companies provided primitive housing for their unmarried workers, but often the bunkhouses offered no comfort. As described by an old-time oil man, one of the oil company camps at Reward, near Taft, contained nothing but "rough board shack[s], full of cracks, furnished with homemade bed frames, a homemade table, chairs and a stove—a piece of twelve-inch stove pipe casing heated with gas." In addition to sleeping accommodations, the camps contained a cookhouse, a clubhouse, an office, and a supply shed. Some of the bunkhouses in the surrounding desert had walls made of burlap hung over wire frames. Water was fed through a trough on the top of the walls and allowed to drip through the burlap to keep it moist. As the wind blew through the wet material, it cooled the inside of the house.

On January 17, 1916, the Midway-Sunset Field was struck by a vicious windstorm. Buildings collapsed, and blowing sheet iron filled the air. Wooden derricks were blown over and smokestacks were toppled. Horses and mules panicked and stampeded through the field. Many workers were injured by falling debris before they could find shelter. Hundreds of rigs were damaged or destroyed. The storm lasted three hours. After it was over, 195 of the 315 derricks at McKittrick needed repairing. Rig builders were summoned from nearby fields to begin the work of rebuilding, and timbers were ordered by the train-car load. Ten days after the first storm, another storm struck. This one missed much of the field, but damaged the nearby towns. Fellows, Taft, and Maricopa were particularly hard-hit. In Maricopa hardly a building escaped. More than a hundred people were left homeless.

A typical oil boom town, Taft lacked proper sanitation facilities, and the unsanitary conditions of course attracted vermin. The situation was compounded by the closeness of the dry Buena Vista Lake bed, which local farmers had converted into a large grainfield, thereby providing a plentiful supply of food for rodents. Generally, most of the mice remained in the grainfields until late 1926 and early 1927, when the region received sufficient rainfall to fill the dry lake bed. Forced from their homes, hordes of mice moved across the countryside toward Taft in search of food.

Millions of rodents overwhelmed the town, "invading homes, stores, warehouses, doghouses, and even gasoline stations," one resident recalled. As food became short, the mice became bolder. To stem the flood of mice, local residents "dug trenches around the drilling rigs and office buildings and poured poisoned grain into them." As many as seventy-five thousand dead mice were counted in a single trench. The smell of the decaying mice was "insufferable," and some physicians feared that an epidemic might be touched off by the carcasses. Cats were useless. After eating their fill, they simply played with the rodents. It was reported that one resident of Taft found sixteen mice playing around her sleeping cat, which was resting after eating its fill of rodents.

It was estimated that between thirty million and one hundred million mice were loose in the field. So many were killed along the road from Taft to Bakersfield that the roadway became slippery. In answer to an appeal for help, the federal government sent to Taft an "extermination expert" named, ironically, Stanley E. Piper. After establishing his headquarters on Pelican Island in the north part of the dry lake bed, he enlisted twenty men, who were promptly dubbed "mouse marines" by local residents, to eradicate the pests. They were helped by the appearance of a large number of birds that fed on the mice. Thanks more to the birds than to Piper's work, during the first six weeks of 1927 the rodent problem disappeared.

As an oil boom town, Taft so captured the public imagination that in 1913 the movie *Opportunity*, starring Fatty Arbuckle, was filmed in the community. During the filming, the producer took great pains to show the excitement and danger of the oil boom. To simulate the taming of a wild well, a dummy derrick was built in the Thirty-Six Hill area, about two miles north of Taft. As the filming started, several hundred gallons of 35¢-per-barrel crude was forced through a pipe so it would appear to be a gusher. Then the oil was set afire, and Arbuckle, who through trick photography was able to remain safely away from the flames, saved the well.

Taft remained in the public limelight when the novel *Polly of the Midway-Sunset* was released in 1917. Written by Janie Chase Michaels, the book purported to be based on life in the community. However, nothing in the fictional account compared to the reality of the "Famous Oil Field Automobile Race of 1912." Scheduled for Washington's Birthday that year, the race offered a first prize of one thousand dollars. Running over a gruelling 106-mile course, the fastest of the racers took more than two and a half hours to complete the circuit. Only three of the thirteen entries finished the required two laps around the circuit—one ended up in a pond after having a blowout. The winner, who covered the 212 miles in 5 hours, 44 minutes, and 59 seconds, was Kern County Deputy Sheriff Jack Bayse. Riding with him as a mechanic was another deputy, Ed Grandy. The following year, on October 27, 1913, Taft's promoters offered a boxing match pitting Sam Langford against Jack Lester. To house the event, a ten-thousand-dollar pavilion seating six thousand fans was constructed on Fourth Street near Recreation Park. Langford won easily, in front of more than five thousand spectators.

Discoveries in the Midway-Sunset Field continued to be made in the late 1920s and early 1930s. On March 3, 1928, the Republic Petroleum Company completed its No. 25 well at a depth of 2,655 feet and opened the producing horizon that became known as the Republic Zone to production. The deeper drilling touched off another round of drilling activity, and by 1941 some twenty additional deep wells had been drilled, with an output ranging from 600 to 2,500 barrels daily.

In the spring of 1935 the Michigan Oil Company began drilling on a well about a mile and a half east of Maricopa, in southwestern Kern County, in what was to become the Gibson Area of the Midway-Sunset Field. In December the Gibson Oil Company acquired the site and completed the well, the Francis No. 1, as a 150-barrel-per-day producer from the Gibson Sand. Encouraged by the strike, Gibson and General Petroleum within the next few months each completed a well in what became the Hoyt Area, about three-quarters of a mile southeast of the Gibson Area. By July 1, 1940, the Gibson Area had produced in excess of 1.3 million barrels of crude, and 98 percent of it came from the Gibson Sand. Production declined rapidly thereafter, and within a short time more than 50 percent of the region's production was saltwater. Even so, natural gas production remained steady at about 2.2 million cubic feet per month.

Midway-Sunset was America's second greatest producer of crude during much of the first half of the twentieth century. Only East Texas had a larger output of oil. Total production from the initial discovery until the end of 1949 amounted to 748,775,000 barrels of oil. For 1949 the field was still the country's fifteenth largest producer.

An early well in the Midway-Sunset Field. Instead of the standard derrick, this well utilized a simple pole affair, supported by guy wires and two smaller poles. Power came from the boiler to the left of the derrick. To its left, a small storage tank had been erected in anticipation of a strike; however, the fact that the boiler was still being used suggests that the well probably had not yet been completed. *Standard Oil Company of California*

Looking up Fourth Street in Taft toward the famous Twenty-Five Hill Field. A typical oil-field boom town, Taft boasted a multitude of false-front, wooden structures along unpaved streets. *American Petroleum Institute*

Construction of a wooden derrick on a cable tool rig at Midway in 1909. The four large supporting beams running from the derrick floor to the crown block were called derrick cornices. The horizontal beams were called derrick girts, and the diagonal braces were called derrick braces. The large square supports under the derrick floor were called foundation posts. *Standard Oil Company of California*

Left: The Lakeview Oil Company's Lakeview No. 1 at Midway shortly after the gusher blew in on March 15, 1910. The Lakeview No. 1 quickly became America's most famous gusher, spewing an estimated ninety thousand barrels of oil daily into the air. *Standard Oil Company of California. Right:* Part of the Lakeview crew attempting to cap the runaway well. They are standing beside one of the twenty large sumps dug in an attempt to save the crude from the wild well. *American Petroleum Institute*

In this panorama of the Lakeview gusher, crude covers the landscape for nearly half a mile around the well. The steam rising at the right of the photograph is from the battery of pumps installed to help control the flow of crude. This is an early view of the runaway, and the belt house can still be seen at the left of the derrick; however, sand has nearly covered the derrick base. *Pacific Oil World*

In the foreground of this photograph is the famous "trout stream," the name given by area oil men to the river of crude flowing from the runaway Lakeview No. 1. *American Petroleum Institute*

In an attempt to contain the crude gushing from the Lakeview No. 1, a sand-bag and sagebrush dam was constructed around the wellhead. *Standard Oil Company of California*

Maricopa, California, one of the boom towns of the Midway-Sunset Field, in 1909. The woman sitting in the chair in the center of the wagon and the man behind her have just been married and are leaving on their honeymoon. Notice the oil-covered clothes of the well-wishers in the back of the wagon and the square telegraph pole. To the left of center in the background are the Oyster House restaurant and the Royal Studio, featuring "photos and views". *Standard Oil Company of California*

Left: Two oil-field workers at Taft diverting a stream of oil from one of the region's wells into a storage pit. Midway-Sunset became well-known for its prolific wells. *Standard Oil Company of California. Right:* The Clendenon Well near Taft. Brought in as a gusher, it flowed at twenty thousand barrels of oil daily. *Henry E. Huntington Library and Art Gallery*

Standard Oil Company of California's gusher in the Midway-Sunset Field in March, 1910. With no place to store the crude, which flowed at ten thousand barrels a day, workers had to allow the oil to run downhill into an earthen pit. *Standard Oil Company of California*

The Standard Oil Company of California's Midway No. 9 burning out of control on January 11, 1911, with a flow of eight thousand barrels of oil daily. To the left of the flames is a shield designed to protect the workers from the heat while they worked to control the fire. *Standard Oil Company of California*

A delivery of casing pipe to a well of the Belridge Oil Company at Midway-Sunset. The wagon-wheeled trailer added to the truck made it long enough to hold the pipe. *Pacific Oil World*

Left: Burning gas well in the Midway-Sunset Field. The onlookers are prudently standing upwind from the flames. *Henry E. Huntington Library and Art Gallery. Right:* Midway gusher, 1912. Notice the tremendous amount of sand thrown out of the wellhead along with the oil. Drifts are clearly visible in the foreground. *Standard Oil Company of California*

A crowd of onlookers attracted to a gusher at Midway-Sunset in 1912. *American Petroleum Institute*

A steam-driven ditcher in operation on the General Petroleum Company's eight-inch pipeline, constructed between Midway-Sunset and San Pedro in 1912. With all the exposed chains and gears, the operator had to be extremely careful not to be pulled into the equipment. The loss of fingers, hands, or arms was not uncommon. The large iron wheels were necessary to support the heavy weight of the machine. *PennWell Publishing Company*

After the ditch had been dug and the pipe lengths joined together, the man in the foreground covered the pipe with "dope," usually creosote, and the others wrapped the pipe with tar paper. This was done both to prevent leaks and to seal the pipe against rust. It was a tedious task that had to be done throughout the entire 154-mile length of the pipeline between Midway-Sunset and San Pedro. *PennWell Publishing Company*

To carry the pipe across this stretch of low land, the Pan American Pipeline Company raised the line onto a series of upright supports and then strung a cable above the pipeline to help support the weight. This particular pipeline ran from Taft to Los Angeles. *PennWell Publishing Company*

A suspension bridge carrying the eight-inch General Petroleum Corporation pipeline across Los Alamos Creek. Such structures were not built according to blueprints, but were "engineered" on the spot by building superintendents. *PennWell Publishing Company*

The handrails and walkways on the Los Alamos Creek suspension bridge allowed the pipe walkers, whose task it was to walk the entire length of the pipeline looking for leaks, to cross the creek. *PennWell Publishing Company*

Pipeline crew laying pipe to a gusher in the Midway-Sunset Field, 1913. *American Petroleum Institute*

An aerial view of the Midway-Sunset Field in 1931. A few of the arroyos have water in them. In a sudden downpour, they could become raging torrents. As a result most of the wells in this picture had been drilled on the crest of the ridges, where they were safe from flash floods. *Standard Oil Company of California*

A Caterpillar tractor cleans up the debris left after a sudden flood down the Santa Clara River after a dam break upstream. This particular flood caused thirty million dollars in property damage and killed four hundred people in 1928. *PennWell Publishing Company*

The remains of a bridge in the Kettleman Hills after a flash flood swept through the area. Sudden floods were a constant danger to oil men drilling in the bottom of arroyos. *Getty Oil Company*

An aerial view looking west from Maricopa into the heart of the Midway-Sunset Field in 1931. Diagonally across the right of the photograph is evidence of a line fight underway between two oil companies that own adjoining leases. As one company would drill a well at the edge of its property line, the other would drill a well just opposite the first to prevent it from draining crude from beneath its lease. Notice the airfield, which appears as an L-shaped figure in the upper left. *Standard Oil Company of California*

This rotary rig was moved from Louisiana to the Midway-Sunset Field in 1908 by the Standard Oil Company of California and proved to be so successful that Standard purchased several new rotary rigs to operate in the pool. *Standard Oil Company of California*

A typical boom-town dwelling near Reward. Built on a slope, the structure had a wooden floor and partial wooden siding. The rest of the walls and the roof were of canvas. While primitive, such modified tents were welcomed by Reward residents, especially the married workers who needed homes for their families. *Getty Oil Company*

Associated Oil Company teams at Reward in 1910. To make this a family portrait, the teamsters have put their wives and children in front of the animals. The team on the left has one hitch of horses in front and one of mules behind them. *Getty Oil Company*

The Associated Oil Company's baseball team at Reward, in 1918. The team has one bat, one ball, and several different types of head gear. *Getty Oil Company*

The main street of Taft, California. While a hotel, several cafes, and a few other businesses have been constructed, several tents are still being used for shelter in the community. On the extreme right is a water wagon. Water was so scarce that it often sold for five cents a glass in the town. To reflect the heat, the water tank has been painted white. *American Petroleum Institute*

The Standard Oil Company of California's Midway Camp in November, 1910. The town of Taft can be seen in the background. Because of the hot weather, covered structures with no walls, such as those just to the left of the large building in the center of the photograph, were built so that the men could work in the shade. *Standard Oil Company of California*

Racing cars were sponsored by several California-based oil companies, and the sport was extremely popular among the state's oil workers. One such race at Taft was dubbed the "Famous Oil Field Automobile Race of 1912." *Atlantic Richfield Company*

A damaged rig after one of the field's vicious windstorms. The wooden derricks were often toppled like sticks when the Santa Ana winds swept through the region. *Getty Oil Company*

Fourth Street in Taft, California, 1910. By this time the community had grown considerably and boasted wooden sidewalks and a multitude of businesses catering to oil-field workers; however, the streets were still unpaved. In the background is the Twenty-Five Hill Field. The single light hanging in the center of the intersection (*top of picture*) was used to summon the police if there was trouble. As the police made their rounds, they kept the light in sight. If a call was received at police headquarters, the light was switched on and the lawmen either hurried back to the station or telephoned for instructions. *Standard Oil Company of California*

An oil well deliberately set afire by Hollywood movie makers to film a motion picture. Notice that instead of having a flaming column of gas shooting from the wellhead, the structure is burning from the ground up. *PennWell Publishing Company*

Elk Hills

ENCOURAGED by the fabulous McKittrick, Kern River, and Midway-Sunset fields, oil men continued to probe the region in search of new pools of petroleum. Eventually, along the eastern flanks of the Temblor Range, they uncovered one of the most controversial of America's oil discoveries—the Elk Hills Oil Field. A long range of low hills, the Elk Hills are about nine miles northeast of Taft and nine miles southeast of McKittrick. Much of the field lay in land given to California by the federal government in 1853 as one of two sections in every township reserved for education.

In 1903, California sold the entire section to Alice Miller for $800. It was later reclaimed by state officials when Miller neglected to pay $3.82 for taxes. It was sold again in 1908, to a man named Hay, for $800. He in turn sold the land to Standard Oil for $12,800. In 1918 Standard Oil spudded in its Hay No. 1 on the property. Completed in mid-January of 1919, it was Elk Hills' first producer and flowed at 200 barrels per day.

Although most California oil men recognized the production possibilities of Elk Hills, development of the region was delayed by the federal government. Acting on a recommendation made by the director of the United States Geological Survey in 1908, federal officials had started in September, 1909, to set aside part of the Elk Hills area as United States Naval Petroleum Reserve No. 1. This initial action was followed on June 26, 1912, by a formal request from the General Naval Board that the secretary of the interior reserve portions of the Elk Hills area. Between August 8 and September 2, 1912, 38,069 acres of the Elk Hills were designated Naval Petroleum Reserve No. 1. Shortly afterward, 29,541 acres in the nearby Buena Vista Hills were named Petroleum Naval Reserve No. 2. As a result of this withdrawal of these 67,610 acres from development, only thirty-five wells were drilled in the entire area prior to 1918—all by the Associated Oil Company, and none commercially successful. Finally the federal government permitted the Standard Oil Company to sink the Hay No. 1 on the land it had acquired, and its success encouraged additional drilling.

Production of natural gas in the Elk Hills Field started when Standard Oil completed its Hay No. 7 on July 26, 1919, in its school-land acreage. The well blew in with a roar that was heard nine miles away in Taft. The huge column of natural gas caught fire from the sparks generated by rocks striking the wellhead and blazed into the sky. Shooting an estimated three hundred feet into the air, the flames could been seen fifty miles away.

The fire was twenty-five feet wide at the base, and it was not safe to approach closer than three hundred yards without protection. Ten boilers were hauled to the well site to provide water, and men hiding behind shields and covered with a spray of water ran lines from them to the well. Even an additional six boilers, though, were not enough to extinguish the flames.

Turning to dynamite, Standard Oil hired Ford Alexander, a well-known firefighter, to blow out the fire with explosives. After stringing a cable from the wellhead to a post some distance away, Alexander placed a trolly on the cable and then loaded the trolly with explosives. He then pulled the trolly along the cable until it was above the wellhead and touched off the dynamite. The explosion had no effect on the flames.

On the tenth day of the inferno, twenty boilers and a more powerful charge of dynamite were used. This time the flames were extinguished. Although the gas continued to spew, the well no longer burned. Eventually the Hay No. 7 became the largest natural gas producer in the world. As of 1927, its output had reached 43 billion cubic feet of natural gas.

In February, 1920, Standard Oil completed a 5,200-barrel-per-day well in the eastern part of the Elk Hills, outside the Naval Petroleum Reserve. Following this second Standard well, development of the eastern portion of the field was rapid. By 1921 the Elk Hills pool was producing 18,085,000 barrels of crude annually—15.8 percent of California's total output.

As nearby development increased, federal officials began to take steps to protect the Naval Reserves. On March 5, 1920, the secretary of the Navy requested legal authority to prevent private companies from drilling next to the Naval Reserves and draining the oil from beneath them. On June 4 of that year Congress passed the necessary legislation. This action, combined with labor unrest, created a period of inactivity in 1922 and 1923.

All efforts to protect the Naval Reserves were negated, though, by the scandal-ridden administration of President Warren G. Harding. In 1921, with the approval of the acting secretary of the Navy, Edwin Denby, President Harding placed the administration of the California Naval Reserves under Secretary of the Interior Albert B. Fall. Between April 25 and December 11, 1922, Fall secretly leased the Elk Hills' reserve to Edward L. Doheny, who, while Fall was secretary of the interior, loaned Fall $100,000 without interest or collateral. After he retired from office, Fall was loaned another $25,000 by Harry F. Sinclair, to whom he previously had secretly leased another Naval Petroleum Reserve in Wyoming.

When the leases became known, a joint congressional resolution charged Fall with fraud and corruption. As a result, on February 28, 1927, the United States Supreme Court invalidated the Elk Hills lease. Later, on June 30, 1927, Fall was indicted for bribery and conspiracy. Convicted of bribery, he was sentenced to a year in prison and fined $100,000. Also indicted for bribery were Doheny and Sinclair. Both were acquitted, but Sinclair was sentenced to nine months in jail and fined $1,000 for contempt of the Senate.

Although the output of the Elk Hills Naval Reserve was curtailed by the legal action, production in the eastern part of the field, outside the reservation boundaries, continued, yielding an average of between 34,000 and 140,000 barrels of crude per acre from 18,690 producing acres. By December 31, 1940, the field had 197 wells that averaged 61 barrels of crude daily. Total production at that time was 154,305,486 barrels of oil. Eventually the total proven acreage in the Elk Hills reached 5,300 acres controlled by private leases and 38,069 acres in the Naval Petroleum Reserves.

Elk Hills was surrounded by several other rich oil and natural gas deposits. In northwest Kern County, along the eastern slope of the Temblor Range, the north area of the Belridge Field was opened to commercial production by the Manel-Minor Petroleum Company in June, 1912. In the late 1920s and early 1930s the deep Temblor Sand was tapped at Belridge. This stimulated deeper drilling in numerous fields in the Southern San Joaquin Valley during the 1930s. The Belridge Oil Company also drilled the discovery well in the South Belridge Area on April 21, 1911. The pool featured relatively highly productive wells—1,200 to 1,500 barrels per day were common—at shallow depths.

The Kettleman Hills, a five-mile-wide and thirty-mile-long stretch of "broken, barren sunbaked" ridges, became a major source of natural gas during the Roaring Twenties, when several new fields were uncovered. The Dudley Ridge Gas Field, southeast of Kettleman City on the southwest edge of Tulare Lake Bed, was located in 1923 by the Tulare Basin Gas Company, Ltd. One well in the field, the Dudley Ridge Syndicate No. 1, blew wild with a flow of 30 million to 40 million cubic feet daily for two weeks.

On November 3, 1926, the Buttonwillow Gas Field was uncovered about thirty miles west and five miles north of Bakersfield. For several years, before the Milham Exploration Company completed its Kern No. 1, oil men had examined the region without success. The discovery well blew in at 4:30 on the afternoon of November 3, 1926, from a depth of 3,323 feet, with so much gas pressure that 400 feet of drill pipe was thrown into the air as the drilling crew rushed to safety. Shortly afterward, the escaping gas was ignited by a spark, and a towering pillar of flame shot into the air. Hearing of the blowout, S. J. Hardison, manager of the Milham Exploration Company, rushed to the scene. Flames that could be seen twenty miles away were shooting 600 feet high when he arrived.

It was estimated that 50 million cubic feet of natural gas a day was burning. Then, at 5:00 A.M. on November 4, the day after the initial blowout, the well caved in and cut off the flow. A crater that the rig disappeared into appeared around the well, and before the area cooled, gas once again began shooting into the air. The roar could be heard for miles, and at 9:49 A.M. it again burst into flames. Not quite an hour later the Kern No. 1

again abruptly stopped flowing. The lull lasted half an hour before the third eruption burst forth. For three days it spewed, building a mound of sand and debris. It did not catch fire again, however, and additional cave-ins sealed the hole.

The Kern No. 1 had uncovered California's first commercial natural gas field. Several other wells were drilled nearby, and by September, 1938, the Buttonwillow Gas Field had twenty-six producing wells. Total production between January 1, 1930, and September, 1938, was 10,115,064,000 cubic feet of natural gas.

Along the border between Kern and Tulare counties, many water wells also produced showings of natural gas. Eventually the Trico Gas Field was uncovered by the Trico Oil and Gas Company on November 19, 1934, about twelve miles west of Delano. About thirty miles northwest of Bakersfield, east of the town of Lost Hills, the Semitropic Gas Field was opened in 1935 by the Standard Oil Company of California.

Numerous other productive pools were discovered in Kern County during the 1920s and 1930s. Among them were the Wheeler Ridge Field, 1923; Round Mountain Field, 1927; Fruitvale Field, 1928; Edison Field, 1934; Mountain View Field, whose various pools were uncovered in 1933, 1934, 1935, and 1939; Wasco Field, 1937; and Strand Field, 1939.

By the mid-1930s, reflective seismographic exploration for oil had proven itself to most California oil men. Based on the principle of measuring underground formations through sound waves sent through the earth by a series of explosions, doodlebugs, as the early seismographs were called, successfully found oil where no surface indications were present. Kern County witnessed some of the earliest seismographic exploration in California, and, as a result, several new discoveries were made in the region. Among them were the Ten Section, named because the Kern County Land Company had previously fenced in a ten-section tract of property in the area. Shell Oil Company's Stevens No. A-1 was spudded in there on January 22, 1936, and completed on June 2, 1936, with an initial flow of 827 barrels daily of oil and 14,123,000 cubic feet of natural gas.

The Rio Bravo Pool, located about fifteen miles northwest of Bakersfield in the flat, alluvium-covered portion of the San Joaquin Valley, also was located by reflection-seismographic exploration. The discovery well, Union Oil Company's Kernco No. 1-34, was spudded in on March 29, 1937. At 10,200 feet company executives considered shutting down the well as a duster. However, the hole was nearing a record-setting depth, and some wanted to see the project through to a conclusion. Although a strong indication of natural gas was found on August 5 between 10,250 and 10,254 feet, no production was located. Finally on November 4, 1937, at a depth of 11,302 feet, the well was completed. During the first four and a half hours, the Kernco No. 1-34 flowed at a rate of 30,000 barrels daily. Afterward the output dropped to 2,400 barrels daily. It was the first well in California to find production at that depth, and at the time the Kernco No. 1-34 was the deepest well in the world.

In 1937 two other pools were opened as a result of a reflection seismographic survey of the region: the Canal Oil Field, to the southwest, and the Greeley Oil Field, west and slightly north of Bakersfield. Eventually two pay zones were defined at Greeley: the Stevens horizon and the Vedder Sand horizon.

At the close of the first four decades of the twentieth century the early development of Elk Hills and its related fields was ending. A period unmatched in the history of the American petroleum industry, it was the final episode in the development of the Kern County fields. Yet Kern County's days in the forefront of the American petroleum industry were far from ended, for the fields in the region have proved to be extremely long-lived, with many of them producing for more than eighty years. Because of the fields' longevity, it has not been unusual to see the heavy wooden derricks built during the early boom years standing beside the permanent steel derricks used decades later, or even more modern portable rigs. In 1982, Kern still ranked first among America's counties in the number of producing wells within its borders.

Left: Standard Oil Company's Hay No. 7 on fire in the Elk Hills, July, 1919. The roar from the well was so loud that it could be heard in Taft, nine miles away. The men standing just behind and to the right of the wellhead are attempting to connect a diversion pipe to control the flames. To protect them from the fire, they are being covered with a spray of water. *Standard Oil Company of California. Right:* A night view of the burning Hay No. 7. Although little but debris remains of the original derrick, men can clearly be seen behind the shields, used to protect them from the heat, as they struggle to bring the flames under control. *Standard Oil Company of California*

A view of the Tupman property in the Elk Hills Field. While most of the pool was within the boundaries of the Naval Petroleum Reserve, there were several rich leases just outside the borders. *Standard Oil Company of California*

Because of the high natural gas pressure, blowouts and fires were common in the Elk Hills Field. This is Standard Oil Company of California's Carman No. 2 blowing wild. Note the rivulets of crude running down the hillside. *Standard Oil Company of California*

Harry F. Sinclair, the third man from the right in this photograph, was acquitted of bribing Secretary of the Interior Albert B. Fall in the Teapot Dome Scandal, associated with opening the Elk Hills Naval Reserve to production. However, he was convicted of contempt of the Senate, fined one thousand dollars, and sentenced to nine months in jail. *Atlantic Richfield Company*

Edward L. Doheny, who was acquitted on technicalities of charges of bribing Secretary of the Interior Albert B. Fall with a $100,000 loan to lease the Elk Hill Naval Reserve to oil men. *Security Pacific National Bank Collection, Los Angeles Public Library*

Secretary of the Interior Ray L. Wilbur inspecting the Milham Exploration Company's Elliott No. 1, in the Kettleman Hills fields in 1929. *Left to right*: Northcutt Ely, executive assistant to the secretary of the interior; R. L. Patterson, U.S. Geological Survey; R. D. Bush, California state oil and gas supervisor; Dr. George O. Smith, chief of the U.S. Geological Survey; and Secretary Wilbur. *PennWell Publishing Company*

Because of the rough terrain, locating a suitable drilling site in the Elk Hills was often difficult. Usually it was easiest to drill in the bottom of one of the many arroyos or to carve out a level area from one of the many small hills, as has been done for the well in the right foreground. *Standard Oil Company of California*

An aerial panorama of the Elk Hills fields in 1931. Notice the orderly, almost geometric, placement of derricks in the pool and the winding maze of roads that twist through the rough countryside to link the wells together. *Standard Oil Company of California*

A later view of the Elk Hills Field. The abundance of small trees that appear in this photograph contrasts with the stark emptiness of early pictures of the field. The area shown here was called "the ghost oil field of Elk Hills," since most of the wells in this photograph were to be plugged and abandoned. *PennWell Publishing Company*

In 1938, the Richfield Company's Tupman Western No. 1 opened the North Coles Levee area to production near the original Elk Hills Field. *Atlantic Richfield Company*

Harold W. Hoots, in the center, the chief geologist for Richfield Oil Company, and A. L. Donneley, the firm's field superintendent, on the right, examining the pit at the Tupman Western No. 1 site after the well opened the North Coles Levee Pool in 1938. *Atlantic Richfield Company*

The Richfield Oil Company of California's first well in the Belridge Field. The tapping of the deep Temblor Sand required the use of steel instead of wooden derricks, since the steel derricks could support the heavier weight necessary to drill deeper wells. *Atlantic Richfield Company*

The Christmas tree on the General Petroleum Company's Berry No. 1 in the Belridge Field. The well was completed on August 11, 1933, at what at the time was the world's record depth—11,377 feet. *PennWell Publishing Company*

One of the Richfield Oil Company's wells in the Belridge Field. Notice the burning of natural gas to the left of the derrick. *Atlantic Richfield Company*

Union Oil Company's Amerada King No. 1 in the Kettleman Hills produced twenty thousand barrels daily. It was drilled at the bottom of an arroyo, and the tops of other nearby derricks can barely be seen over the crest of the ridge. *Union Oil Company of California*

A group of lease hounds in the Kettleman Hills checking on the progress of a well. A lease hound's job was to report back to his company all the activity of competing oil companies in his area. Notice that most of the men are wearing wide-brim straw hats for protection from the hot sun. *Atlantic Richfield Company*

A series of stairs and walkways on this Richfield Oil Company well in the Kettleman Hills allowed the drilling crew easy access to the derrick floor in spite of the rough terrain. *Atlantic Richfield Company*

Drill floor of a rotary rig operating in the Kettleman Hills. The two men in the rear are standing clear while the man on the left "runs up kelly." The kelly, or grief stem, was the heavy square pipe that worked through the square home in the rotary table to drive the drill stem. Its was called a grief stem because "it brings grief to a worker who is struck by it in making a connection, and to a driller who makes an unsuccessful boring." *Atlantic Richfield Company*

Much of the drilling in the Kettleman Hills was done by rotary rigs. This is a photograph of the vibrating rotary mud screen in operation on a Richfield Oil Company rig in the pool. As the rotary bit drills downward, mud is forced down through the pipe and back up the hole, carrying the bit cuttings with it to the surface. So that the mud can be recycled, it is emptied onto a vibrating screen that separates the mud from the cuttings, which are pulled toward the lower edge of the screen in this photograph. The cuttings are then disposed of as waste and the mud channeled back into the sump for re-use. *Atlantic Richfield Company*

View of a rotary rig operation in the Kettleman Hills fields. Instead of using a bit to pound its way through the earth, a rotary rig drilled its way down. The bit was attached to numerous stems of pipe, and the pipe was rotated by the turntable, shown in the foreground in this photograph. *Atlantic Richfield Company*

Left: Richfield Oil Company's Signal No. 1 in the Kettleman Hills fields. Concrete and wooden restraining walls have been constructed around the base of the drilling rig to provide a foundation for the derrick in the sandy soil. *Atlantic Richfield Company. Right:* The Milham Exploration Company's Kern No. 1 in the Buttonwillow Field. Known for the volume of its natural gas production, the pool became a major supplier of natural gas in the San Francisco Bay area. *PennWell Publishing Company*

The Wheeler Ridge Field, about twenty-five miles southwest of Bakersfield. Notice how the drilling site and the connecting roads in the field had to be carved out of the sides of the hills. The smaller, steeper cuts at the right in the photograph carry pipelines. *American Petroleum Institute*

Because of the tremendous natural gas production in the southern end of the San Joaquin Valley in the Kettleman Hills, the Pacific Gas and Electric Company connected the region with San Francisco Bay by a system of sixteen- and twenty-two-inch pipes in 1929. These men are preparing to unload the next joint of pipe, to be attached to the one barely visible on the left. When the pipeline was completed, it was 250 miles long, which, at that time, made it the longest pipeline on the West Coast. *PennWell Publishing Company*

Another portion of the Pacific Gas and Electric Company's pipeline network tying the gas-producing area of the Kettleman Hills to the rest of California. The pipe is being welded electrically. As the hoist on the tractor lifts the pipe into place, the man on the left, who is wearing a round shield to protect his eyes, welds the sections together. Apparently this is a marshy area, because several boards have been placed on the ground so that the Caterpillar tracks will not sink into the ground. *PennWell Publishing Company*

The Pacific Gas and Electric Company's Kettleman Hills compressor plant used to push the natural gas through its extensive pipeline network. The equipment is powered by five 750-horsepower Cooper gas engines. *PennWell Publishing Company*

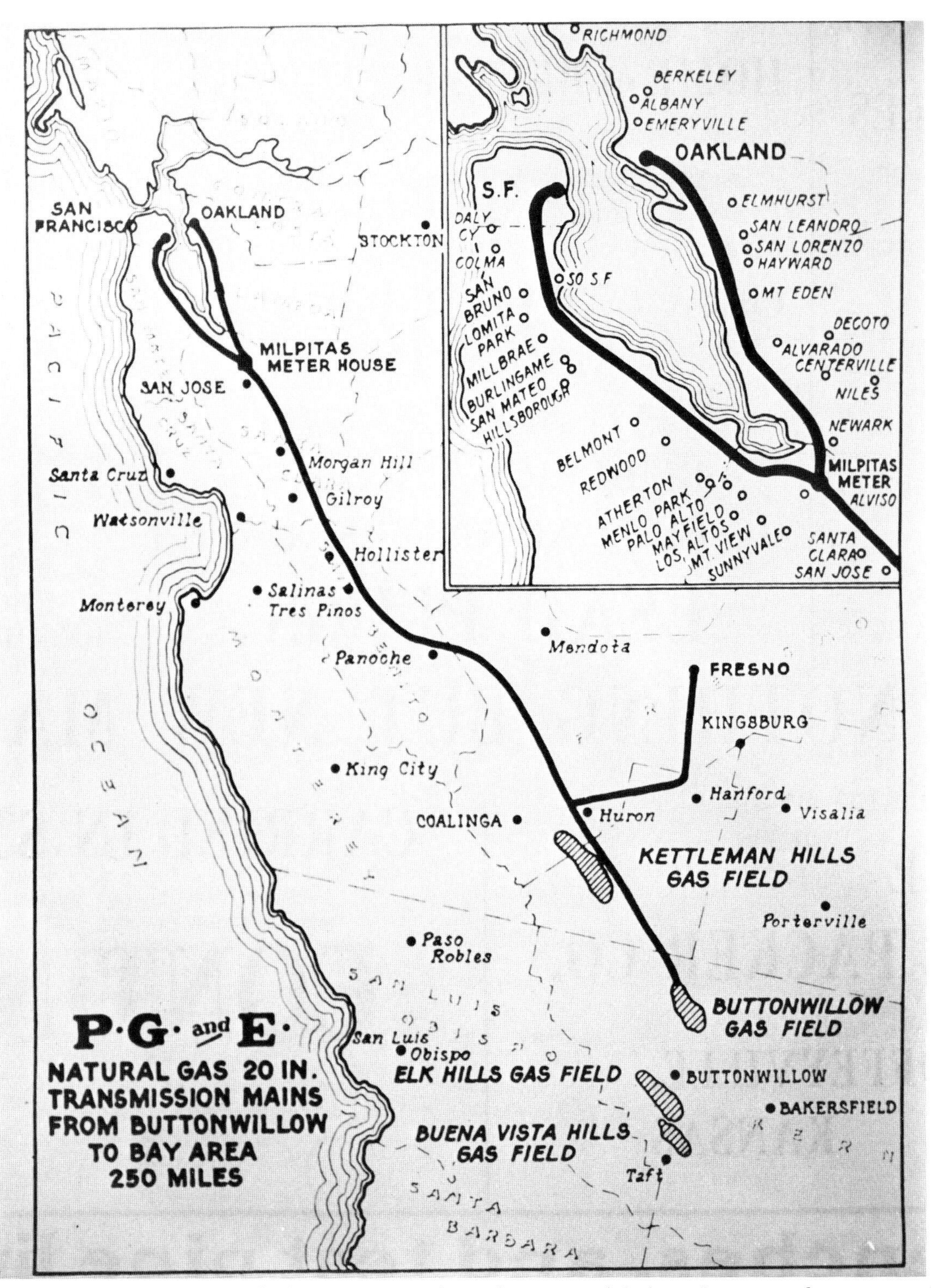

A map of the Pacific Gas and Electric Company's pipeline network linking the gas-producing area of the Kettleman Hills to the markets in the San Francisco Bay area. *PennWell Publishing Company*

Epilog

BETWEEN 1900 and 1929, the California petroleum industry reached its zenith. Between 1901 and 1940 the state ranked first among America's oil-producing states in fourteen years, second in twenty-one years, and third in three years. By 1929 annual production in the state reached 292 million barrels of crude—the largest ever for a single year. Cumulative production at that time totaled 3,308,368,000 barrels. Afterward production began to decline as the market was saturated with crude.

ANNUAL OIL AND NATURAL GAS PRODUCTION AND VALUE

Year	*Petroleum**	*Value*	*Natural Gas†*	*Value*
total through 1875	175,000	$472,500		
1876	12,000	30,000		
1877	13,000	29,250		
1878	15,227	30,454		
1879	19,858	39,716		
1880	40,552	60,828		
1881	99,862	124,828		
1882	128,636	252,272		
1883	142,857	285,714		
1884	262,000	655,000		
1885	325,000	750,750		
1886	377,145	870,205		
1887	678,572	1,357,144		
1888	690,333	1,380,666	12,000	$10,000
1889	303,220	368,048	14,500	12,680
1890	307,360	384,200	41,250	33,000
1891	323,600	401,264	39,000	30,000

* number of barrels
† thousands of cubic feet

Year	*Petroleum**	*Value*	*Natural Gas†*	*Value*
1892	385,049	561,333	75,000	55,000
1893	470,179	608,092	84,000	68,500
1894	783,078	1,064,521	85,080	79,072
1895	1,245,339	1,000,238	110,800	112,000
1896	1,257,780	1,180,793	131,100	111,457
1897	1,911,569	1,918,269	71,300	62,657
1898	2,249,088	2,376,420	111,165	74,424
1899	2,677,875	2,660,793	115,110	95,000
1900	4,329,950	4,152,928	40,566	34,578
1901	7,710,315	2,961,102	120,800	92,034
1902	14,356,910	4,692,189	120,968	99,443
1903	24,340,839	7,313,271	120,134	75,237
1904	29,736,003	8,317,809	144,437	91,035
1905	34,275,701	9,007,820	148,345	102,479
1906	32,624,000	9,238,020	168,175	109,489
1907	40,311,171	16,783,943	169,991	114,759
1908	48,306,910	26,566,181	842,883	474,584
1909	58,191,723	32,398,187	1,148,467	616,932
1910	77,687,568	37,689,542	10,579,933	1,676,367
1911	84,648,157	40,552,088	5,000,000	491,859
1912	89,689,250	41,868,344	12,600,000	940,076
1913	98,494,532	48,578,014	14,210,836	1,053,292
1914	102,881,907	47,487,109	16,529,963	1,049,470
1915	91,146,620	43,503,837	21,992,892	1,706,480
1916	90,262,557	57,421,334	28,134,365	2,871,751
1917	95,396,309	86,976,209	44,343,020	2,964,922
1918	99,731,177	127,459,221	46,373,052	3,289,524
1919	101,182,962	142,610,563	52,173,503	4,041,217
1920	103,377,361	178,394,937	58,567,772	3,898,286
1921	112,599,860	203,138,225	67,043,797	4,704,678
1922	138,468,222	173,381,265	103,628,027	6,990,030
1923	262,875,690	242,731,309	240,405,397	15,661,443
1924	228,933,474	274,652,874	209,021,596	15,153,140
1925	232,492,147	330,609,829	194,719,924	15,890,082
1926	224,673,281	345,546,677	214,549,477	19,465,347
1927	231,195,774	260,735,498	224,668,940	20,447,294
1928	231,811,465	229,998,680	260,887,116	22,260,947
1929	292,534,221	321,366,863	342,214,000	29,067,546
1930	227,328,988	271,699,046	315,513,952	24,559,840
1931	188,270,605	141,835,723	344,959,920	16,690,695

Year	*Petroleum**	*Value*	*Natural Gas†*	*Value*
1932	177,745,286	142,890,247	284,168,872	16,272,061
1933	172,139,362	143,063,972	271,743,544	15,403,514
1934	174,721,282	159,529,671	263,207,517	14,408,761
1935	205,979,855	179,335,311	302,447,193	17,680,661
1936	214,733,315	211,667,185	298,922,708	18,585,970
1937	238,558,562	237,845,872	323,883,714	19,859,865
1938	249,395,763	258,354,343	332,358,843	22,310,755

The value of California's cumulative production of gold as of 1938 stood at $2,060,925,706, while the combined value of all the oil and natural gas produced in the state as of the same date was $5,463,764,404. Between 1900 and 1937, California produced 16.45 percent of the world's oil output. During this same period the Long Beach Field produced 2.15 percent of the world's oil, Santa Fe Springs 1.56 percent, Coalinga 1.24 percent, Kern River 1.17 percent, and Huntington Beach 1.00 percent.

By the end of the first half of the twentieth century, California had produced twenty-one of America's top eighty-one oil fields, based on cumulative output. In 1949 three California fields were in the top ten American fields based on annual production: Wilmington was second in the entire nation, Ventura Avenue was fifth, and Coalinga Nose was ninth. In that year Buena Vista still ranked thirteenth, Midway-Sunset fifteenth, and Kettleman North Dome eighteenth.

In January, 1982, California ranked fourth—behind Texas, Louisiana, and Alaska—among America's richest petroleum-producing states. Its output of 33,627,000 barrels of crude for that month was 7.01 percent of the nation's total production. California ranked second, behind only Louisiana, in offshore production for the first month of 1982, with more than one-sixth of the nation's offshore production.

Potential offshore production was in the forefront of California's oil industry in the early 1980s. Of particular interest to oil men was the Santa Maria Basin just off and north of Point Conception. By mid-1982, Chevron, USA, a subsidiary of Standard Oil Company of California, and its partner in the venture, Phillips Petroleum Company, were reporting a discovery that had the potential to produce 160,000 barrels of oil daily in the Point Arguello–Hueso area. In addition, Texaco, Inc., was reporting similar test results in the Santa Maria Basin, an area that some oil-industry analysts claimed would prove to be the largest strike made in the United States in fifteen years.

The discovery well was drilled approximately eleven miles offshore from Point Conception, about forty miles north of the Santa Barbara Channel blowout and spill of 1969. Two other promising structures, Point Arguello and Hueso are about a mile and a half to the west. In addition to Texaco, other oil companies involved in the search include Atlantic Richfield Company, Union Oil Company of California, Getty Oil Company,

Pennzoil, and Conoco, Inc. By October, 1982, the various companies had spent $2.7 billion exploring the region. Estimates of reserves in the area vary between five hundred million and one billion barrels of crude. The story of California oil—now unfolding largely in the waters of the Pacific off the state's coast—should continue indefinitely, yielding energy for the nation, wealth for some, controversy, and drama.

Left: During World War II, many women entered the California petroleum industry, as more and more oil-field workers left for military service. By the time the war was over, the women had made a permanent place for themselves. This worker is measuring the amount of oil in a steel storage tank. *Standard Oil Company of California. Right:* Because so many of California's oil fields were located in developed areas, oil wells lived side by side with other businesses and homes throughout the state. To accommodate their neighbors, California oil companies covered their drilling rigs with soundproofing to lessen noise pollution. In addition, the covering prevented oil from spilling onto neighboring property should a blowout occur. This well is being drilled on a car lot in Culver City. *PennWell Publishing Company*

The soundproofing of wells was improved so much in California that by the mid-1960s wells literally were being drilled in people's backyards with little inconvenience. This well is being drilled in September, 1968, between a residence and a mobile home park without disrupting the life-styles of those living nearby. *Standard Oil Company of California*

Drilling platforms in the Santa Barbara Channel in 1968. Special precautions, including installing additional blowout preventers, have been taken to prevent a repeat of the disastrous Santa Barbara spill. Two wells are being sunk from the same platform in the foreground, while a single derrick is operating from the well in the background. The drilling rig on the center platform is just being erected. *Cities Service Oil Company*

To lessen the chances of spills into the ocean, wells that are drilled offshore on platforms are connected by pipeline to onshore service facilities. This tank battery and other equipment service a joint well of the Phillips Petroleum Company and Cities Service Oil Company in the Santa Barbara Channel. Notice the dike around the storage tank on the left. The depression the tank is in is large enough to contain the entire contents of the tank without letting any oil leak into the ocean. *Cities Service Oil Company*

Bibliography

American Petroleum Institute. *Petroleum and Figures, Ninth Edition, 1950.* New York: American Petroleum Institute, 1951.

Atwood, Albert W. "When the Oil Flood Is On." *Saturday Evening Post*, July 7, 1923, p. 3.

Beaton, Kendall. *Enterprise in Oil: A History of Shell in the United States.* New York: Appleton-Century-Crofts, 1957.

Beringer, Pierre N. "Coalinga: The Great Oil Producing Field of California." *Overland Monthly* 53 (February, 1909): 156–70.

Boone, Lalia Phipps. *The Petroleum Dictionary.* Norman: University of Oklahoma Press, 1952.

"California Depression." *Scientific American*, January, 1950, p. 30.

Crafts, H. A. "California's Newest Oil Fields." *Overland Monthly* 44 (November, 1904): 489–92.

David, Alfred H. "The Fight for Oil." *Overland Monthly* 60 (September, 1912): 284–88.

"Detailed Statistics." *Petroleum Supply Monthly*, May, 1982, pp. 35–73.

Ellison, O. C. "Kern City and the Kern River Oil Districts." *Overland Monthly* 38 (July, 1901): 66–90.

Franks, Kenny A. *The Oklahoma Petroleum Industry.* Norman: University of Oklahoma Press, 1980.

———, and Paul F. Lambert. *Early Louisiana and Arkansas Oil: A Photographic History, 1901–1946.* College Station: Texas A&M University Press, 1982.

Getty, J. Paul. *My Life and Fortunes.* New York: Duel, Sloan and Pearce, 1963.

———. *As I See It: The Autobiography of J. Paul Getty.* Englewood, N.J.: Prentice-Hall, 1976.

Griswold, H. T. "California's Oil Boom." *Current Literature* 30 (February, 1901): 167–68.

"Hand Pumps and Tin Cans Start Miniature Oil Boom." *Popular Science Monthly*, December, 1935, p. 24.

Hewins, Ralph. *The Richest American: J. Paul Getty.* New York: E. P. Dutton & Co., 1960.

Hutchinson, W. H. *California: Two Centuries of Man, Land and Growth in the Golden State.* Palo Alto, Calif.: American West Publishing Company, 1969.

———. *Oil, Land and Politics: The California Career of Thomas Robert Bard.* 2 vols. Norman: University of Oklahoma Press, 1965.

Jenkins, Olaf P., ed. "Geological Formations and the Economic Development of the Oil Fields of California." *Bulletin No. 118, California Department of Natural Resources, Division of Mines.* April, 1943.

Kempton, Dwight. "Drilling Submarine Oil Wells as Performed at Summerland, Cal.," *Scientific American*, January 18, 1902, p. 36.

Lockwood, Charles. "In the Los Angeles Oil Boom, Derricks Sprouted Like Trees." *Smithsonian*, October, 1980, pp. 187–206.

Lowenstein, Roger. "Firms Hopeful on Oil Drilling off California." *Wall Street Journal*, October 14, 1982.

Marcosson, Isaac F. "The Black Golconda." *Saturday Evening Post*, April 5, 1924, p. 12.

Morris, Richard B., ed. *Encyclopedia of American History*. New York: Harper and Row, 1976.

Nadeau, Remi. *Los Angeles: From Mission to Modern City*. New York: Longmans, Green, and Co., 1960.

Noble, Earl B. "Forty-Two Years of Service with the Union Oil Company Ends for W. W. Orcutt, when Death Closes His Career." *Union Oil Company Bulletin*, May, 1942, pp. 1–3.

Orcutt, W. W. "History: Early Days in the California Oil Fields." Archives, Union Oil Company of California, Los Angeles.

Rister, Carl C. *Oil! Titan of the Southwest*. Norman: University of Oklahoma Press, 1949.

Rintoul, William. *Drilling Ahead: Tapping California's Richest Oil Fields*. Santa Cruz, Calif.: Valley Publishers, 1981.

———. *Oildorado: Boom Times on the West Side*. Fresno, Calif.: Valley Publishers, 1978.

———. *Spudding In: Recollections of Pioneer Days in the California Oil Fields*. San Francisco: California Historical Society, 1975.

Sainz, Darwin. "Oil—Its Contribution to the Santa Maria Valley." Archives, Union Oil Company of California, Los Angeles.

Smith, Bertha H. "Unusual People: California's Petroleum Queen." *Illustrated World*, September, 1916, pp. 7–72.

Stalder, Walter. "Memoir of a Distinguished Life Member of the American Society of Civil Engineers." *Union Oil Company Bulletin*, June, 1942, pp. 9–24.

Struth, H. J., ed. *The Petroleum Data Book, 1947*. Dallas: Petroleum Engineer Publishing Co., 1947.

Union Oil Company of California. *Sign of the 76: The Fabulous Life and Times of the Union Oil Company of California*. Los Angeles: Union Oil Company of Calif., 1977.

Warren, Herbert O. "California's First Oil Well Is Still Producing." *Scientific American*, July, 1928, pp. 60–61.

Wells, W. B. *Wild Bill the Driller*. Hoffman Printing Co., 1977.

White, Gerald T. *Formative Years in the Far West: A History of Standard Oil Company in California and Predecessors through 1919*. New York: Appleton-Century-Crofts, 1962.

Wilbur, Ray L. "Gold of All World's Mines Dwarfed by Oil Field." *Popular Mechanics*, September, 1931, pp. 427–29.

Woehlke, W. V. "California's Black Gold." *Sunset*, August, 1910, pp. 173–87.

Index